Rare Earth Elements
Essential Materials for Energy and Sustainable Development

Editors:

Vipin Kumar Sharma
Professor and Head
Department of Mechanical Engineering
Meerut Institute of Engineering and Technology
Meerut, India

Mohamed Arezki Mellal
Professor
LMSS, Department of Mechanical Engineering
Faculty of Technology, M'Hamed Bougara University
Boumerdes, Algeria

CRC Press
Taylor & Francis Group
Boca Raton London New York

CRC Press is an imprint of the
Taylor & Francis Group, an **informa** business
A SCIENCE PUBLISHERS BOOK

First edition published 2026
by CRC Press
2385 NW Executive Center Drive, Suite 320, Boca Raton FL 33431

and by CRC Press
4 Park Square, Milton Park, Abingdon, Oxon, OX14 4RN

CRC Press is an imprint of Taylor & Francis Group, LLC

Library of Congress Cataloging-in-Publication Data (applied for)

ISBN: 978-1-041-10808-5 (hbk)
ISBN: 978-1-041-16191-2 (pbk)
ISBN: 978-1-003-68325-4 (ebk)

DOI: 10.1201/9781003683254

Typeset in Times New Roman
by Prime Publishing Services

Preface

The unique properties of rare earth metals, which enhance the functionality and efficiency of various technologies, are the primary emphasis of this book's discussion of their application in the energy sector. Important objectives include improving battery performance, offering sustainable energy solutions, raising energy efficiency, and developing renewable energy technologies. The book *Rare Earths Elements: Essential Materials for Energy and Sustainable Development* covers in great detail the development and optimization of systems for dealing with clean energy technologies and their applicability to a variety of sectors. Academics and students interested in studying sustainable manufacturing systems and clean technology will find these articles useful. The book comprises fourteen chapters.

Vipin Kumar Sharma and Mohamed Arezki Mellal

Foreword

Rare earth elements (REEs), referred to as the "seed of technology", are widely employed in numerous advanced technologies, including low-carbon technologies, namely, in electric vehicle motors and wind turbine generators. These elements are also utilized in hydrogen electrolysers, energy-efficient fluorescent lighting, and permanent magnets, which are crucial components of wind turbines. As the globe switches to renewable energy, there will likely be a rise in the need for REEs. It is anticipated that the energy sector will dominate the mineral markets.

Over the next ten years, it is anticipated that both the demand and the supply of these vital minerals will increase significantly due to the global shift away from fossil fuel energy and toward low-carbon energy. REEs are necessary for conventional energy as well, but the combination of these elements that is crucial to energy in the future is different from what was required in the past.

Historically, the catalytic qualities of REEs have been their main application in industrial processes. Their role in dissolving larger molecules into smaller hydrocarbons that can be used as fuels is crucial to the petrochemical industry. The market for REEs has grown significantly in the recent past. The primary difficulties will be clarified in the final section: diversifying and addressing social and environmental issues with rare earth mining to guarantee a steady supply in the future. The book *Rare Earths Elements: Essential Materials for Energy and Sustainable Development* brings researchers and industrial parts to the latest works in this topic.

Vipin Kumar Sharma and Mohamed Arezki Mellal

Contents

List of Contributors

Ashish Kumar
Assistant Professor, Faculty of Management & Commerce, Poornima University, Jaipur, Rajasthan, India.

Atul Pratap Singh
Department of Pharmacy, IIMT University, Meerut, Uttar Pradesh, India.

Avitaj Kaushal
Department of Biotechnology, Shobhit University, Meerut, U.P., India.

Bhaskar Garg
Department of Chemistry, School of Basic Sciences and Technology, IIMT University, Meerut 250001, Uttar Pradesh, India

Darshan Srivastav
Department of Applied Science & Humanities, Maharana Pratap Institute of Technology, Gorakhpur, Uttar Pradesh, India.

Devika Agarwal
Professor, Faculty of Management & Commerce, Poornima University, Jaipur, Rajasthan, India.

Dhananjay Taumar
Department of Pharmacy, IIMT University, Meerut, Uttar Pradesh, India.

Dhowmya Bhatt
Department of Computer Science Engineering, Faculty of Engineering and Technology, SRM Institute of Science and Technology, Delhi NCR Campus, Ghaziabad, UP, India (201204).

Dinesh Kumar
Department of Mechanical Engineering, MMEC, Maharishi Markandeswar (Deemed to be University), Mullana-Ambala, India.

Disha Mishra
Department of Applied Science & Humanities, Maharana Pratap Institute of Technology, Gorakhpur, Uttar Pradesh, India.

Garima Bartariya
Associate Professor, School of Life Science and Technology, IIMT University, Meerut, UP, India.

Gourav Kumar Sharma
Department of Mechanical Engineering, NIT, Jalandhar, Punjab, 144011, India.

Gurmeen Rakhra
Department of Biochemistry, School of Bioengineering and Biosciences, Lovely Professional University, Phagwara,144411, India.
Department of Mechanical Engineering, Maharishi Markandeshwar (Deemed to be) University, Mullana, Ambala, India.

Gurseen Rakhra
Department of Nutrition & Dietetics, School of Allied Health Sciences, Manav Rachna International Institute of Research and Studies, Faridabad, Haryana-121004, India.

Himanchal Sharma
Department of Pharmacy, IIMT University, Meerut, Uttar Pradesh, India.

Juhie Agarwal
Assistant Professor, Vardhman PG College, Bijnor, UP, India.

Mahesh Kumar Gupta
Department of Mechanical Engineering, Faculty of Engineering and Technology, SRM Institute of Science.

Mahiti Gupta
Department of Bio-Sciences & Technology, MMEC, Maharishi Markandeswar (Deemed to be University), Mullana-Ambala, India.

Manpreet Singh
Assistant Professor, Department of Mechanical Engineering, Baba Farid College of Engineering and Technology, Punjab-151002, India.
Applied Science Research Centre, Applied Science Private University, 11931, Amman, Jordan.
Assistant Professor, Department of Mechanical Engineering, Chitkara University Institute of Engineering and Technology, Chitkara University, Rajpura-140401, Punjab, India.

Mohamed Arezki Mellal
LMSS, Department of Mechanical Engineering, Faculty of Technology, M'Hamed Bougara University Boumerdes, Algeria.

Pardeep Kumar
Department of Mechanical Engineering, MMEC, Maharishi Markandeswar (Deemed to be University), Mullana-Ambala, India.

Prachi Bhatnagar
Department of Computer Science & Engineering, IIMT University, Meerut (UP), India.

Puneet Kumar
Associate Professor, School of Management and Commerce, IIMT University, Meerut Uttar Pradesh, India.

Rakhi Gupta
Department of Computer Science & Engineering, IIMT University, Meerut (UP), India.

Ranjana Singh
Department of Commerce and Management, IIMT University, Meerut (UP), India.

Riyad Abu-Malouh
Department of Mechanical and Industrial Engineering, Applied Science Private University, Amman 11931, Jordan.

Sangeeta Sharma
Associate Professor, School of Life Science and Technology, IIMT University, Meerut, UP, India.

Sanjeev Kumar Sharma
Department of Applied Science & Humanities, Maharana Pratap Institute of Technology, Gorakhpur, Uttar Pradesh, India.

Sanjukta Vidyant
Department of Biotechnology, Shobhit University, Meerut, U.P., India.

Santosh Kumar Rai
Department of Mechanical Engineering, Faculty of Engineering and Technology, SRM Institute of Science and Technology, Delhi NCR Campus, Ghaziabad, Uttar Pradesh, India (201204).

Shahanawaj Ahamad
College of Computer Science and Engineering, University of Hail, Saudi Arabia.

Shallender Bhati
Department of Mechanical Engineering, Meerut Institute of Engineering and Technology, Meerut, U.P., 250001, India.

Sharad Kumar
Department of Biotechnology, Shobhit University, Meerut, U.P., India. School of Engineering and Technology Shri Venkateshwara University, Gajraula, Amroha, U.P., India.

Shivali Rathore
Assistant Professor, Dept. of Botany, Govt. K.R.G. PG Autonomous College, Gwalior, MP, India.

Soniya Goyal
Department of Biosciences and Technology, Maharishi Markandeshwar (Deemed to be) University, Mullana, Ambala.

Subhash Chand
Department of Mechanical Engineering, IIMT University, Meerut (UP), India.

Sudhir Agrawal
Department of Applied Science & Humanities, Maharana Pratap Institute of Technology, Gorakhpur, Uttar Pradesh, India.

Tanuja Bisht
Department of Chemistry, Indira Priyadarshini Government Girls P. G. College of Commerce, Haldwani 263139, Uttarakhand, India

Tarun Kumar
Vashishth School of Computer Science and Applications, IIMT University, Meerut, U.P., India.

Vatan Chaudhary
Department of Pharmacy, IIMT University, Meerut, Uttar Pradesh, India.

Vikas Goyat
Department of Mechanical Engineering, Faculty of Engineering and Technology, SRM Institute of Science.

Vikas Sharma
Department of Computer Applications, SRM Institute of Science and Technology, Delhi NCR Campus Ghaziabad, U.P., India.

Vipin Kumar Sharma
Department of Mechanical Engineering, Meerut Institute of Engineering and Technology, Meerut, Uttar Pradesh, India.

Yong-Chien Ling
Department of Chemistry, National Tsing Hua University, 101, Section 2, Kuang-Fu Road, Hsinchu 30013, Taiwan

1 A Review of Sustainable Recovery of Rare Earth Elements: Bridging Resource Security and Circular Economy Goals

Disha Mishra[1*], Darshan Srivastav[2], Sanjeev Kumar Sharma[3], Sudhir Agrawal[4], Vipin Kumar Sharma[5] and Mohamed Arezki Mellal[6]

1. Introduction

With the rising tide for sustainable development and low carbon economies, the demand for green technology and materials suitable for green technology is on surge. In this picture, the Rare Earth Elements (REEs) hold an important place. The unique magnetic, catalytic, and luminescent properties possessed by these elements make them suitable for a variety of applications across industries. Some of the crucial applications include manufacturing of wind turbines, motors of electric vehicles (EVs), energy-efficient lighting, and high-performance electronics, making REEs an important pillar for clean-tech innovation (Massari & Ruberti, 2013). Regardless of their abundance in the Earth's crust, their concentration is not at par with that required for economic mining. As such, the extraction of REEs have become both challenging and environmentally taxing.

The escalating climate crisis, economic interdependence, and unsustainable practices in mining have aggravated the interest in alternative methods for deriving REEs. Some of these methods include the use of secondary sources like electronic waste, residues from industry, and end-of-life green technologies for recovery and recycling of REEs (Binnemans et al., 2013). Exploration of these alternatives is a thrust area, as it reduces the dependence on primary mining ensuring environmentally viable mining practices and prolonged availability of these critical resources.

Moreover, amidst the prevailing geopolitical tensions, the growing global demand for REEs and green technology is posing significant concerns about the supply security of these elements, as most of the production and processing

[1-4] Department of Applied Science & Humanities, Maharana Pratap Institute of Technology, Gorakhpur, Uttar Pradesh, India.

[5] Department of Mechanical Engineering, Meerut Institute of Engineering and Technology, Meerut, Uttar Pradesh, India.

[6] LMSS, Faculty of Technology, M'Hamed Bougara University, Boumerdes, Algeria.

* Corresponding author: dishamishra106@gmail.com

DOI: 10.1201/9781003683254-1

capacity of REEs is attributed to China, with over 60% of the global production (Humphries, 2013). As a result, apart from the technological aspects, there is also an imperative dive across the nations for regulatory frameworks and sustainable models for circular economy and supply of REEs.

1.1 Importance of Rare Earth Elements in the Green Economy

REEs are pivotal to bring about the transition for green energy. Neodymium and dysprosium are used to make permanent magnets in wind turbines to increase the efficiency (U.S. Department of Energy, 2011). In EVs, REEs are used to improve the performance of electric motors and are also found in nickel-metal hydride batteries (Golev et al., 2014). Europium and terbium are critical in manufacturing of LED and LCD displays, while lanthanum and cerium are used in catalytic converters, and petroleum and refinery industry.

The availability and sustainable management of REEs is a critical parameter that reflects a nation's potential to meet the climate goals. For instance, owing to its high supply risk, The European Union, for instance, has identified REEs as critical raw materials of economic importance (European Commission, 2023).

In India, the Ministry of Mines (Govt. of India) is working expeditiously to ensure supply chain resilience for critical metals including REEs. In this regard, the Government has taken several initiatives like exploration projects for augmenting REE resources in the country; collaborating with nations like Brazil and Dominican Republic for developing cooperation in the field of Rare Earth Minerals and Critical Minerals (Press Information Bureau, 2025).

This chapter presents a comprehensive overview of the sustainable recovery of REEs, with emphasis on methods and techniques that are congruent with the circular economy principles and environmental sustainability. Specifically, it addresses the following objectives:

- To examine the environmental and geopolitical challenges associated with primary REE mining;
- To explore potential secondary sources of REEs, including e-waste and industrial residues;
- To evaluate emerging recovery technologies, including bio-based and AI-enabled methods;
- To analyse the role of circular economy strategies in enhancing REE supply chain resilience;
- To highlight current policy initiatives, regulatory frameworks, and industrial practices supporting REE recovery.

The scope of this chapter encompasses both technological and policy perspectives, aiming to bridge the gap between academic research and practical application. It incorporates recent advances in green chemistry, environmental engineering, and digital technologies to propose a roadmap for sustainable REE management in alignment with the goals of climate mitigation, economic resilience, and technological self-reliance.

2. Rare Earth Elements: Properties and Applications

2.1 Classification and Chemical Characteristics

REEs refer to a group of 17 chemically similar elements comprising the 15 lanthanides (from lanthanum to lutetium) along with scandium and yttrium. These elements occupy the f-block of the periodic table and share common chemical properties due to the filling of 4f orbitals (Gupta & Krishnamurthy, 2005). Although the term 'rare' suggests scarcity, REEs are relatively abundant in the Earth's crust. For instance, cerium is more abundant than copper. However, their occurrence in dispersed form and the challenge of economically viable concentration make them rare in terms of accessibility (Castor & Hedrick, 2006).

REEs are broadly classified into two groups based on atomic number and geochemical behaviour:

- **Light Rare Earth Elements (LREEs):** La, Ce, Pr, Nd, Sm
- **Heavy Rare Earth Elements (HREEs):** Gd, Tb, Dy, Ho, Er, Tm, Yb, Lu, along with Y and Sc (though not lanthanides, they exhibit similar chemical behaviour).

HREEs are typically rarer and more valuable due to their limited supply and specific industrial uses. These elements exhibit unique magnetic, optical, and catalytic properties, making them indispensable for modern high-tech applications (Zepf, 2013).

2.2 Industrial and Technological Applications

REEs are vital in various industries owing to their diverse functional characteristics. Some of the key application areas included in Figure 1.

2.2.1 Energy Sector

Permanent magnets made from neodymium (Nd), praseodymium (Pr), dysprosium (Dy), and terbium (Tb) are widely used in wind turbine generators and EV motors due to their high magnetic strength and thermal stability. The transition to low-carbon energy systems has significantly increased the demand for these magnets (Global E-Waste Monitor, 2024).

2.2.2 Electronics and Communication

Europium (Eu) and terbium (Tb) are essential in producing red and green phosphors for fluorescent and LED displays. Yttrium (Y) is used in color television tubes and LED lighting. Lanthanum (La) and cerium (Ce) are utilized in polishing powders for optical lenses and glass manufacturing (Golev et al., 2014).

2.2.3 Automotive Industry

REEs are used in catalytic converters to reduce emissions from internal combustion engines. Cerium and lanthanum are also used in fuel additives and metal alloys

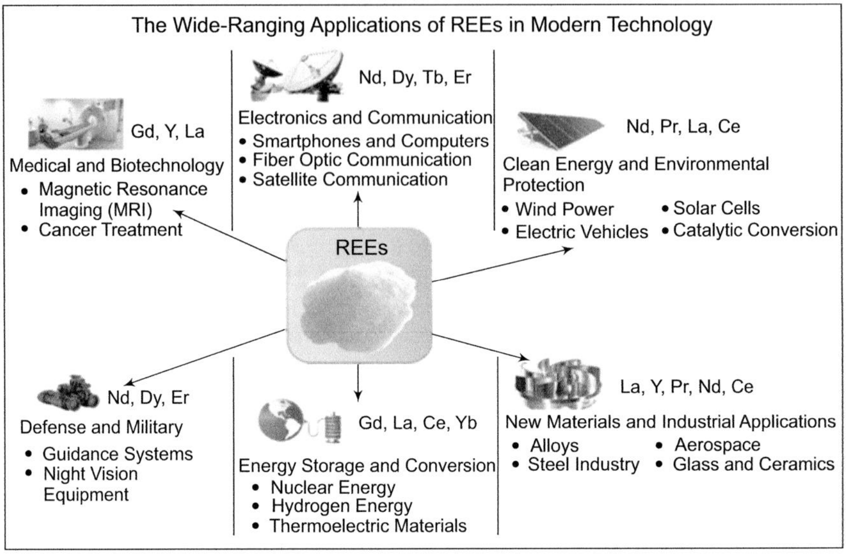

Fig. 1 Application Areas of REEs in modern technology (Balaram, 2019, Sharma et al., 2019, Patel et al., 2024, Patel et al., 2025, Sharma et al., 2024).

to enhance fuel efficiency and engine performance (Jordens, Cheng, & Waters, 2013; Srikant et al., 2021; Patel et al., 2025; Patel et al., 2025; Sehgal et al., 2025).

2.2.4 Medical Applications

Gadolinium (Gd) is widely used as a contrast agent in magnetic resonance imaging (MRI) due to its paramagnetic properties. Samarium-153 is used in cancer treatment, especially for relieving bone pain in metastatic cancers (Xie et al., 2021).

2.2.5 Aerospace and Defence

Rare earths contribute to high-performance alloys, sensors, radar systems, and laser rangefinders. They are used in jet engines and military technologies where heat resistance and precision are critical (Hatch, 2012; Sharma et al., 2019; Sharma et al., 2019; Sharma et al., 2019; Sharma et al., 2020; Sharma et al., 2021).

2.3. Role of REEs in Energy, Electronics, and Mobility Sectors

The role of REEs in the energy transition is irreplaceable. As the world moves towards cleaner and more sustainable sources of energy, REEs facilitate this transformation through their presence in wind turbines and e-mobility both will be driven by technological advancements and optimization of material usage and by the political ambitions underlying their development, whereas demand in other sectors, such as for electronics and specialized equipment, will mostly be influenced by market dynamics. This gives rise to a wide range of possible future scenarios (Figure 2).

- **Wind Turbines:** Nd-Fe-B magnets in direct-drive wind turbines enhance efficiency and reduce mechanical complexity (Alonso et al., 2012).
- **Electric Vehicles:** EVs use REE-based magnets in motors and batteries. Hybrid vehicles use nickel-metal hydride batteries that contain lanthanum and cerium (US DOE, 2011; Kumar et al., 2022; Patel et al., 2024).
- **Smart Devices:** Smartphones, tablets, and other digital devices depend on REEs for miniaturized, energy-efficient performance. For example, yttrium and europium are used in backlighting and displays.

Given these crucial applications, the demand for REEs is projected to rise significantly in the coming decades. According to the International Energy Agency (IEA, 2024), the demand for rare earths could increase by three to seven times by 2040 depending on the scale of the global clean energy transition.

3. Challenges in Primary Mining of Rare Earth Elements (REEs)

While REEs are critical to the advancement of green technologies and digital infrastructure, their extraction through primary mining is fraught with numerous environmental, geopolitical, and socioeconomic challenges. These challenges not only raise questions about the sustainability of current supply chains but also highlight the urgent need for alternative sourcing methods such as recycling and urban mining.

3.1 Environmental Impacts of Conventional Mining

Out of the various industrial activities that are environmentally intensive, one of the most challenging areas is the extraction and processing of REEs. Rare earth ores, such as bastnaesite, monazite, and xenotime exploit chemical processes like acid leaching, solvent extraction, and high-temperature treatments, which generate hazardous waste such as radioactive tailings and acidic effluents, in significant amounts (Golev et al., 2014).

The radioactive nature of thorium and uranium, which form the constituent of monazite ores pose significant environmental concerns. Countries like China and Malaysia have already suffered problems of contamination of soil and groundwater near the mining regions, which is the outcome of improper disposal of radioactive tailings (Ali, 2014). Also, the acid-drainage mining sites lead to long-term ecological damage in surrounding ecosystems.

Moreover, REE mining also causes greenhouse gas emissions. According to reports, production of one ton of REEs may be accompanied by creation of up to 2,000 tonnes of toxic waste, and an exhaustive amount of fossil fuel energy, thereby escalating the climate change (Rim et al., 2013).

3.2 Geopolitical and Supply Chain Risks

The global supply of REEs is highly concentrated. As of 2022, China accounted for over 60% of global REE production and more than 85% of processing

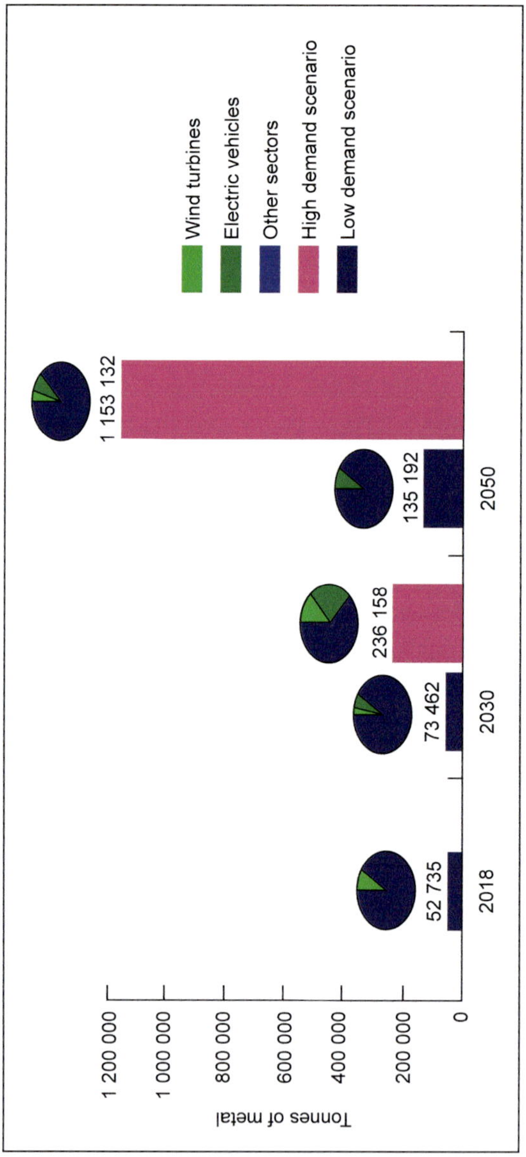

Fig. 2 Estimated global demand of REEs from wind turbine generators, EV motors and other sectors (Alves Dias, Bobba, et. al., 2020).

capacity, giving it near-monopolistic control over the market (USGS, 2023). This concentration has significant geopolitical implications, as nations dependent on REE imports face risks associated with trade disputes, export restrictions, and political instability.

The 2010 diplomatic dispute between China and Japan, where China temporarily halted REE exports to Japan, served as a wake-up call for many industrial nations. Since then, countries like the United States, Australia, and members of the European Union have sought to diversify supply chains and invest in domestic sources (Humphries, 2013).

China stands as the global epicenter for REEs, dominating both production and reserves in 2024. The left chart (Figure 3) reveals that China accounts for a staggering 69% of worldwide REE mine output, far surpassing the U.S., Myanmar, Australia, and other countries. Meanwhile, the right chart highlights that China also holds nearly half (49%) of all known world reserves of REEs, with Brazil, India, Australia, Russia, and others considerably behind. This concentration underscores China's pivotal influence over the global REE supply chain—an influence with profound implications for industries and nations dependent on these critical materials for technologies central to the clean energy transition and advanced electronics.

3.3 Socioeconomic Limitations and Community Impacts

Several REE mining sites are located in rural and underdeveloped regions. Weak or inappropriate regulatory framework in such areas result in socioeconomic adversities. For instance, problems like occupational hazards, child labour, and human rights abuse, due to informal and artisanal mining practices are significant in African countries (Leung et al., 2021).

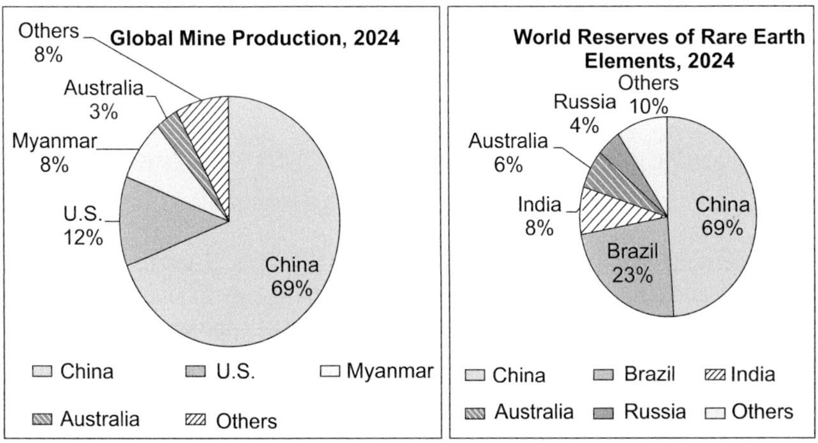

Fig. 3 The supply and reserves of REEs are fragmentally shared, with mostly concentrated with China, accounting for 69% of global production and 49% of world reserves, as of 2024. U.S., Myanmar, Brazil, India, Australia, and Russia are among the other significant players.

As a result of exposure to toxic substances, several communities near mining and processing facilities in China have reported health issues including waterborne illness, cancer, and respiratory diseases (Wübbeke, 2013).

Many times, local communities have to face displacement, loss of livelihoods, disruption in their environment and ecosystem, etc., which often result in social unrest and lack of trust on industrial activities. These issues aggravate the problems resisting the operation and long-term viability of mining activities.

3.4 Economic Viability and Resource Exhaustion

Amidst the increasing demand for REE, the economic viability of primary mining is subject to multiple constraints like market fluctuations, high capital investment, and the complicated isolation process of co-occurring elements. REE occurs in the form of mixed deposits and the isolation process requires multiple steps which involve high costs (Jordens et al., 2013).

Furthermore, many high-grade REE reserves are undergoing rapid depletion, and exploration of new deposits is subject to regulatory frameworks, environmental scrutiny, and public resistance. Ore qualities have declined while the cost of extraction is on surge. All these factors necessitates the efforts for secondary recovery technologies.

4. Secondary Sources of Rare Earth Elements (REEs) and Their Potential

Secondary sources refer to the materials that contain recoverable REEs but are not obtained directly from mining. These sources present significant potential for bridging the supply–demand gap. Examples of these sources include electronic waste (e-waste), industrial residues, and end-of-life products from green technologies. Recovery from these sources is both economically beneficial and environmentally responsible, aligning with circular economy goals (Binnemans et al., 2013).

4.1 Secondary Sources of REE

(i) Electronic Waste (E-waste)

One of the richest yet least utilized sources of FREEs is the E-waste. The magnets, phosphors, and circuit boards found in devices such as smartphones, laptops, hard drives, fluorescent lamps, and televisions, contain REEs in very small quantities which are much valuable for recovery purposes.

REE Distribution in E-waste

- **NdFeB Magnets:** These magnets present in hard drives, speakers, and headphones, contain elements like neodymium (Nd), praseodymium (Pr), and dysprosium (Dy).

- **Phosphor Powders:** The powder obtained from compact fluorescent lamps (CFLs) and display screens, contain europium (Eu), terbium (Tb), and yttrium (Y).
- **Battery Components:** Elements such as lanthanum (La) and cerium (Ce) are used in Nickel-metal hydride (NiMH) based batteries.

Utilizing E-waste for recovery of REEs may result in multifaceted benefits, such as – reduction in greenhouse gas emissions, mitigation of toxic waste, etc. According to a survey by the United Nations, the global generation of E-waste is on the rise with a rate of 2.6 million tonnes annually. This increase may amount to a total generation of 82 million tonnes of E-waste by 2030. The global generation of E-waste is rising five times faster than the rate of recycling (Global E-waste Monitor, 2024). This highlights both the impending need for recycling of e-waste and also the huge opportunity for resource recovery.

Apart from the modern methods of REE recovery from E-waste, such as *hydrometallurgy*, *pyrolysis*, and *bioleaching*; the recently developed AI-enabled sorting systems are also very helpful in segregating REE-rich components before processing (Yang et al., 2022).

(ii) Industrial Waste

Large volumes of REE-rich waste are generated in industry processes such as production of aluminium, coal combustion, and phosphate fertilizer manufacturing.

(iii) Red Mud (Bauxite Residue)

Red mud is obtained as a by-product in the process of aluminium extraction, known as Bayers Process. It contains significant quantities of REEs like cerium (Ce), lanthanum (La), and neodymium (Nd), with concentrations ranging from 200 to 1,500 ppm (Ochsenkühn-Petropoulou et al., 2002). In order to exploit these sources, development of recovery processes including acid leaching and membrane separation, are under development.

(iv) Coal Fly Ash

REEs with composition of around 500 ppm is present in fly ash obtained as by combustion product from burning of coal. Yttrium (Y), cerium (Ce), and lanthanum (La) are examples of such elements found in fly ash (Taggart et al., 2018). Coal is still being used as the basis of power production in many countries, this indicates a valuable yet underutilized REE source.

(v) Phosphogypsum

Due to the geochemical similarity between phosphate and REE-bearing minerals, phosphogypsum also contains REEs. Phosphogypsum is a waste product obtained during the process of phosphate fertilizer production. Methods like bioleaching and solvent extraction can be exploited for extraction of REEs from this source (Borai et al., 2020).

4.2 End-of-Life Green Technologies

Technologies that contribute to green energy such as wind turbines, EVs, and solar panels are themselves rich in REEs at the end of their useful life. Recovering REEs from these systems helps sustain clean energy supply chains and supports circular resource loops.

4.2.1 Wind Turbine Magnets

Direct-drive wind turbines use permanent magnets made of Nd, Pr, Dy, and Tb. These components are non-replaceable during operation but become excellent sources for REE recovery after decommissioning (Widmer et al., 2015).

4.2.2 Electric Vehicle Batteries and Motors

Hybrid and EVs contain REEs in traction motors and battery packs. NiMH batteries use La and Ce, while motors contain Nd and Dy magnets. By 2030, retired EVs are expected to become a significant secondary resource base for REEs (Schreiber et al., 2016).

4.2.3 Fluorescent and LED Lamps

Used lighting systems contain REEs in phosphor coatings. Though the amount per unit is small, the large quantity of waste bulbs globally represents a substantial cumulative resource (Zeng et al., 2016).

4.3 Urban Mining and Circular Resource Flows

Urban mining—the process of reclaiming raw materials from spent products, buildings, and waste—offers a sustainable path to REE recovery. Modern cities are repositories of valuable resources embedded in obsolete devices and infrastructures (Brunner, 2011). Urban mining:

- Reduces landfill pressure;
- Lowers energy input vs. virgin mining;
- Enhances material security for industrial nations.

Integrating urban mining with formal recycling sectors can transform waste into wealth while reducing environmental footprints. Figure 4 summarizes the main extraction techniques for REEs. An overview of secondary sources of REEs, as discussed in this section is presented in Table 1.

5. Sustainable Recovery Techniques of Rare Earth Elements (REEs)

To address the environmental concerns, supply risks, and economic limitations of primary mining, researchers and industries have shifted focus toward sustainable and efficient recovery methods for REEs. These techniques, often applied to

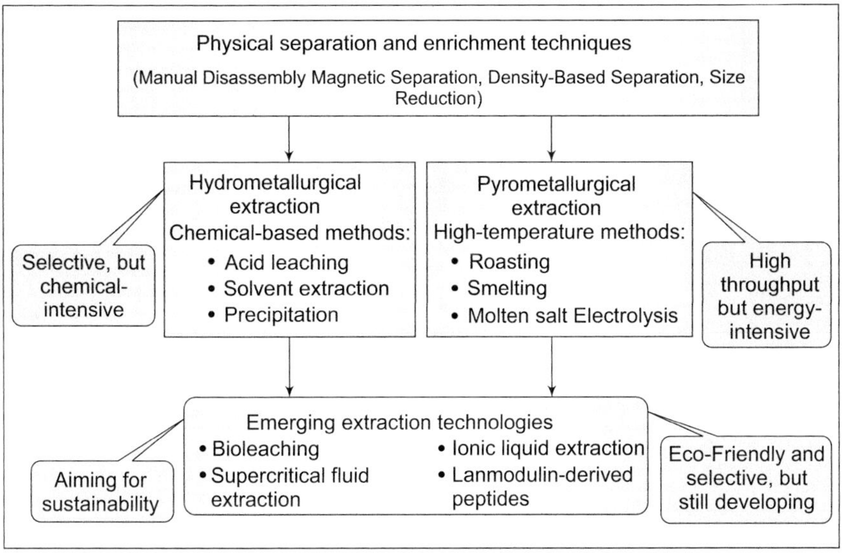

Fig. 4 Overview of REE recovery methods from e-waste: physical separation, hydrometallurgical extraction, pyrometallurgical extraction, and emerging green technologies (Balaram, 2019).

Source: "Rare earth elements: A review of applications, occurrence, exploration, analysis, recycling, and environmental impact" by V. Balaram, published in Geoscience Frontiers (2019).

secondary sources such as e-waste, industrial residues, and end-of-life green technologies, offer eco-friendly and scalable alternatives. This section explores various emerging and established technologies for sustainable REE recovery.

5.1 Hydrometallurgical Methods

Hydrometallurgy involves leaching REEs from solid materials using aqueous solvents, typically acids or bases, followed by separation and purification.

5.1.1 Acid Leaching

Sulfuric acid (H_2SO_4), hydrochloric acid (HCl), and nitric acid (HNO_3) are commonly used to dissolve REEs from ores, fly ash, and red mud (Abaka-Wood et al., 2021). The leachate is then processed using precipitation, solvent extraction, or ion exchange.

5.1.2 Alkaline Leaching

Particularly suitable for ion-adsorption clays, alkaline leaching uses ammonium salts (e.g., ammonium sulfate) to extract REEs under mild conditions (Xie et al., 2021).

Table 1 Overview of secondary sources of REEs.

Secondary Source	*Key Materials Containing REEs*	*Advantages*	*Challenges*	*Examples/Applications*
E-waste and End-of-Life Products	Permanent magnets, phosphors, batteries, catalysts	High REE content; urban mining potential	Low collection rates, informal recycling, technical recovery issues	NdFeB magnets from HDDs, fluorescent lamps
Industrial Waste and By-products	Coal fly ash, red mud (bauxite residue), phosphogypsum	Large volume availability; can valorise waste	Low REE concentration; presence of impurities	Red mud in aluminium industry; fly ash in coal power plants
Spent Catalysts and Magnets	Automotive and petrochemical catalysts; industrial magnets	High selectivity and REE content in spent components	Deactivation complexity; contamination with other metals	Fluid catalytic cracking (FCC) units in refineries
Renewable Energy and Green Tech Waste	Wind turbine magnets, EV motors, LED phosphors	Future waste stream; supports green tech circularity	Complex multi-material structure; long lifespan products	End-of-life wind turbines; hybrid vehicle traction motors

Although efficient, traditional hydrometallurgy has drawbacks, including large volumes of hazardous liquid waste and high reagent costs. New trends focus on **closed-loop leaching** systems and **green solvents** like ionic liquids and deep eutectic solvents (DESs), which offer high selectivity and lower toxicity (Zhang et al., 2022).

5.2 Pyrometallurgical Methods

Pyrometallurgy involves high-temperature processes such as roasting, smelting, and sintering to extract REEs. These methods are mainly applied to magnet alloys (e.g., NdFeB) and metallic residues.

- **Carbothermic and metallothermic reduction:** Reduce REE oxides to metallic form.
- **Selective roasting:** Used to convert REE-bearing phases into more leachable forms.

Although thermally robust and effective, pyrometallurgy is energy-intensive and can emit greenhouse gases (Binnemans et al., 2013). Therefore, it is often combined with hydrometallurgy in hybrid processing chains.

5.3 Bio-based Extraction (Biometallurgy)

Bio-based extraction, or biometallurgy, uses living organisms like bacteria, fungi, and plants, to leach, bind, or accumulate REEs from waste materials.

5.3.1 Bioleaching

Acidophilic microorganisms like *Acidithiobacillus ferrooxidans* and *Leptospirillum ferrooxidans* oxidize sulphur or iron, producing acids that solubilize REEs from ores or e-waste (Nancharaiah & Mohan, 2022).

5.3.2 Biosorption

Microbial cell walls, especially in fungi and algae, contain functional groups (e.g., carboxyl, hydroxyl) that can adsorb REEs from aqueous solutions. *Aspergillus niger* and *Saccharomyces cerevisiae* have demonstrated high biosorption capacity (Wang et al., 2021).

5.3.3 Phytomining

Certain hyperaccumulator plants can uptake REEs from soil and concentrate them in biomass. After harvesting, the plants are incinerated, and the ash is processed to extract metals (López et al., 2019).

Bio-based techniques offer low energy requirements, minimal emissions, and adaptability to low-grade ores. However, limitations include long processing times and sensitivity to environmental conditions.

5.4 Solvent Extraction and Ion Exchange

Solvent extraction (SX) remains the dominant industrial method for separating individual REEs due to their chemical similarity.

- **Organophosphorus-based extractants** like D2EHPA and PC88A selectively separate REEs based on ionic radius and charge density.
- **Deep eutectic solvents (DESs)**, a newer class of eco-friendly solvents, offer high REE selectivity, low volatility, and easy recyclability (Sheng et al., 2020).

Ion exchange resins are also employed in tandem with leaching systems to capture REEs from solutions with high purity and minimal loss. While solvent extraction requires careful handling of organic phases, it is scalable and widely used in both primary and secondary REE processing (Abrahami et al., 2015).

5.5 AI-Enabled Sorting and Smart Recovery Systems

Artificial Intelligence (AI) and automation are revolutionizing the identification, classification, and separation of REE-containing components from complex waste streams.

5.5.1 Optical and X-ray Sorting

Advanced imaging systems, such as X-ray fluorescence (XRF) and near-infrared (NIR), can detect REE-rich components in electronic scrap and segregate them in real-time (Yang et al., 2022).

5.5.2 Robotic Disassembly

AI-integrated robotic systems are capable of disassembling electronic devices and isolating high-value REE components like magnets and phosphor powders with minimal human intervention (Tan et al., 2021).

5.5.3 Machine Learning in Process Optimization

AI algorithms optimize process conditions for leaching, solvent extraction, and waste minimization, enabling smart decision-making in REE recycling plants (Gautam et al., 2022).

AI-enabled systems increase recovery efficiency, reduce labour intensity, and improve the safety and speed of recycling operations, especially in high-throughput facilities. An overview of green extraction and recovery techniques for REEs is presented in Table 2.

6. Role of Circular Economy in REE Management

The circular economy (CE) offers a transformative framework for managing REEs sustainably. Unlike the traditional linear economy, which follows a "take-make-dispose" model, the CE emphasizes **resource efficiency, product longevity, reuse, remanufacturing, and recycling** to minimize waste and environmental degradation (Geissdoerfer et al., 2017). In the context of REEs, CE principles are essential to reduce dependency on primary extraction and to maintain secure, clean, and cost-effective material flows in green technology supply chains.

6.1 Principles of Circular Economy Applied to REEs

Circular economy in REE management revolves around three core principles:

- Designing out waste and pollution
- Keeping products and materials in use
- Regenerating natural systems (Ellen MacArthur Foundation, 2015)

Applying these principles to REEs means redesigning products for disassembly and reuse, developing systems for recovery from end-of-life products, and enhancing the durability and reparability of REE-containing technologies.

6.2 Eco-Design and Product Life Extension

Product eco-design involves developing electronics, motors, batteries, and other REE-dependent technologies in a way that facilitates easier disassembly, component recovery, and recyclability (Bocken et al., 2016). This approach ensures that REEs embedded in complex devices can be efficiently reclaimed at the product's end-of-life.

Examples include:

- Modular design in smartphones for easier extraction of REE-based components
- Magnet-to-magnet recycling in motors and generators without complete material breakdown (Walton et al., 2015).

By extending the functional lifespan of REE-intensive products, eco-design contributes significantly to circularity and material efficiency.

6.3 Urban Mining and Reverse Logistics

Urban mining refers to the recovery of valuable materials like REEs from urban waste streams, including electronic waste, construction debris, and decommissioned green technologies (Brunner, 2011). With rising urbanization and e-waste generation, cities are becoming significant repositories of secondary resources.

To enable urban mining, **reverse logistics systems** must be established, including:

- Collection centres for obsolete electronics
- Take-back schemes by producers or retailers
- Incentivized recycling programmes for consumers.

Studies show that urban mining can yield higher REE concentrations than some primary ores, making it a promising and sustainable alternative (UNEP, 2013).

Table 2 Overview of Green Extraction and Recovery Techniques for REEs.

Technique	*Principle*	*Advantages*	*Limitations*	*Applications / Use Cases*
Hydrometallurgy	Leaching REEs from ores or wastes using aqueous solutions (acids, solvents)	High recovery efficiency; suitable for low-grade ores	Generates liquid waste; often uses toxic chemicals like HCl or HNO_3	Extraction from e-waste, red mud, coal ash
Pyrometallurgy	Uses high-temperature smelting or roasting to extract REEs	Effective for hard-to-leach materials; less chemical use	High energy consumption; not suitable for all REE compounds	Processing of permanent magnets and industrial scrap
Biometallurgy	Utilizes microorganisms (fungi, bacteria) to bioleach or bioaccumulate REEs	Eco-friendly; low energy requirement; minimal chemical usage	Slow process; influenced by pH, temperature, and bioavailability	Recovery from fly ash, spent catalysts, low-grade ores
Solvent Extraction (SX)	Separates REEs based on their differential solubility in organic solvents	High selectivity; commercially proven for light and heavy REEs	Uses hazardous organic solvents; multi-stage process	Separation of REE mixtures in hydrometallurgical flowsheets
Ion Exchange & Adsorption	Uses resins or materials to adsorb REEs from solution	Reusable materials; minimal chemical input	Limited to dilute solutions; capacity depends on material surface area	Recovery from acid mine drainage or electronic leachates
Deep Eutectic Solvents (DES)	Novel green solvents replacing traditional acids/organics	Low toxicity; biodegradable; tunable properties	Still under development; cost and scalability remain concerns	Selective REE leaching from phosphors and spent magnets
AI-Enabled Sorting Systems	AI and sensor-based systems for automated material separation	Enhances efficiency and accuracy; reduces manual labor	High upfront cost; requires digitized infrastructure	Smart sorting of REE-containing e-waste and battery components

6.4 Product-as-a-Service (PaaS) and Extended Producer Responsibility (EPR)

PaaS models promote the idea that manufacturers retain ownership of products and lease them to customers. This ensures the return of products for remanufacturing or material recovery, including REEs (Stahel, 2016). Examples include leasing EV batteries or wind turbines, ensuring their recovery and reuse at the end-of-life phase.

The linear approach follows a "take, make, use, dispose" process resulting in waste, while the CE emphasizes continuous reuse, repair, collection, and recycling of materials, aiming to minimize waste and maximize resource efficiency (Figure 5).

Extended Producer Responsibility (EPR) is a policy approach wherein producers are made responsible for the entire lifecycle of their products, especially disposal and recycling. EPR programmes have been implemented in the European Union and Japan to ensure effective collection and processing of REE-containing devices (OECD, 2016). The image (Figure 6) illustrates the circular process of EPR in plastic waste management, showing the interconnected roles of producers, importers, brand owners, consumers, and waste processors in ensuring that plastic packaging is systematically collected, recycled, and responsibly disposed of to reduce environmental impact.

6.5 Benefits and Implementation Challenges

Benefits:

- Reduced reliance on primary mining and imports;
- Lower environmental impact and carbon footprint;
- Increased supply chain resilience;
- Creation of green jobs and local value chains.

Challenges:

- Lack of infrastructure for waste collection and recycling
- Technological limitations in material separation and recovery
- Policy gaps and lack of global standardization
- Low public awareness and participation.

Successful implementation of CE strategies for REEs will require a **multi-stakeholder approach** involving industry, government, academia, and civil society.

7. Global Policy and Regulatory Landscape

A robust and coordinated policy framework is crucial to promoting the sustainable recovery, recycling, and responsible sourcing of REEs. Given the strategic importance of REEs for clean energy, defence, and digital technologies, many governments have developed national strategies, international collaborations,

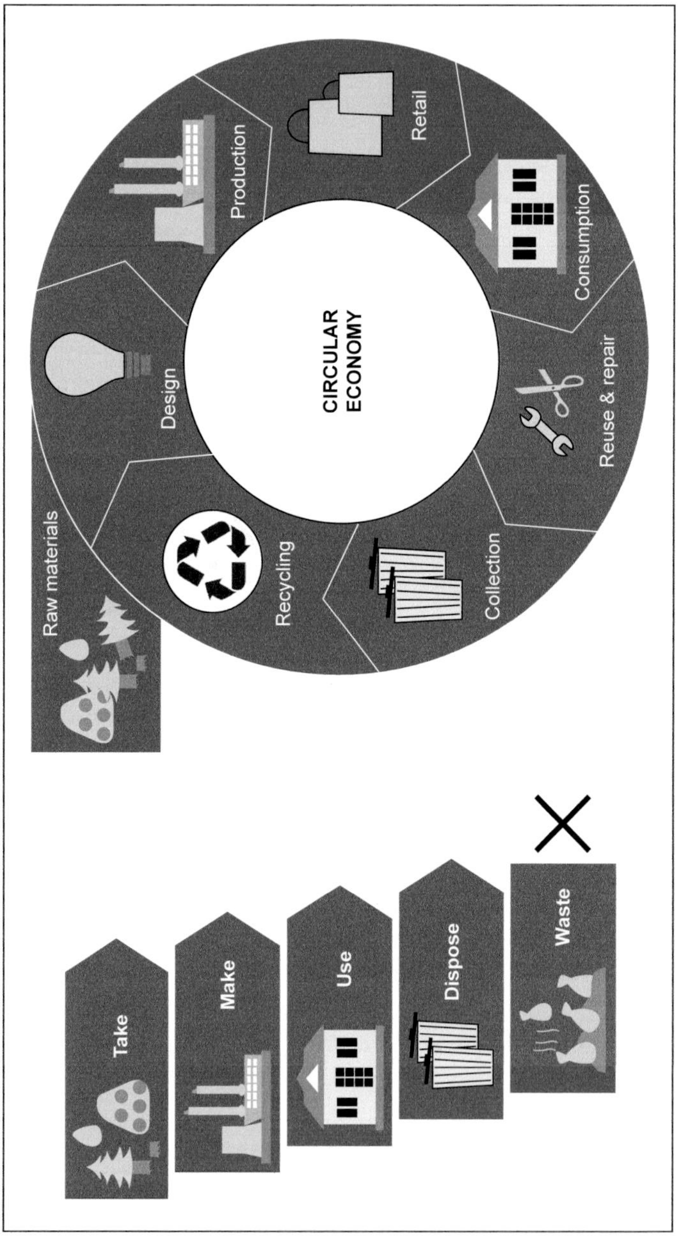

Fig. 5 Comparison of linear and circular economy models.

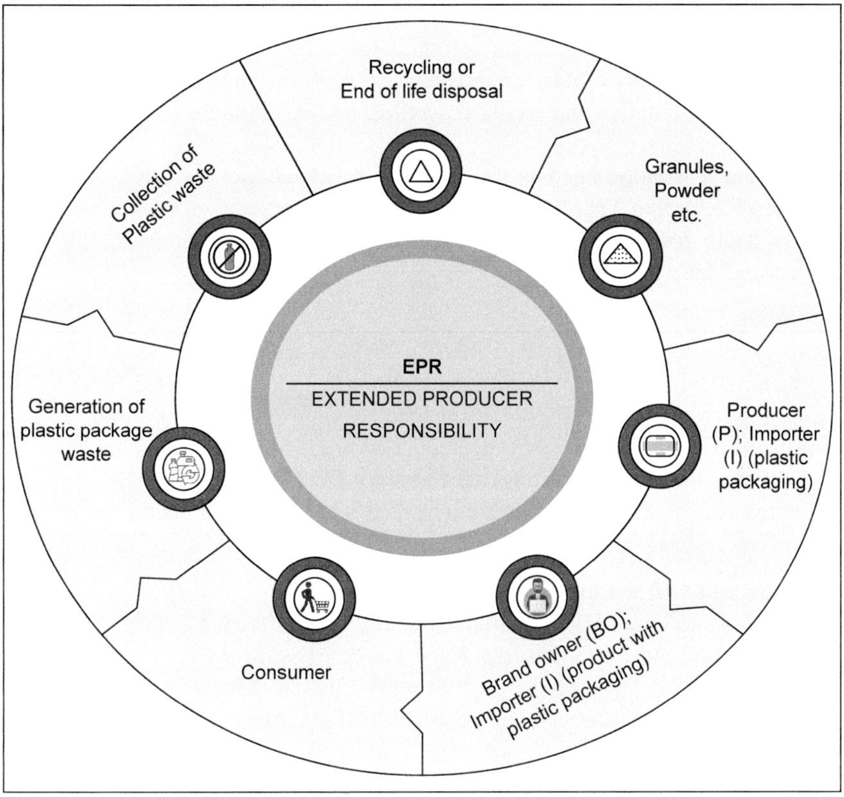

Fig. 6 Overview of the Extended Producer Responsibility (EPR) framework (Courtesy: Central Pollution Control Board (CPCB) of India).

and regulatory mechanisms to ensure secure and sustainable supply chains. This section explores these global efforts and evaluates the role of public and private stakeholders in advancing REE recovery.

7.1 International Initiatives and Strategic Frameworks

(a) The European Union (EU)

The European Commission has identified REEs as **critical raw materials** due to their high supply risk and economic significance. The **EU Critical Raw Materials Act (2023)** outlines plans to diversify supply sources, support research in recycling technologies, and strengthen partnerships with resource-rich countries (European Commission, 2023).

The **European Raw Materials Alliance (ERMA)** fosters industrial collaboration and funding for sustainable REE projects across the EU. Furthermore, the EU mandates **eco-design and extended producer responsibility** through directives like WEEE (Waste Electrical and Electronic Equipment), helping to recover REEs from e-waste streams (Schüler et al., 2011).

(b) The United States

The U.S. has elevated REEs as essential for national security. The **Critical Materials Institute (CMI)**, established by the Department of Energy, funds research on alternative materials, recycling, and sustainable mining practices (DOE, 2022).

Recent legislation such as the **Bipartisan Infrastructure Law (2021)** and **Inflation Reduction Act (2022)** allocates billions toward critical mineral recovery, supply chain resilience, and clean technology infrastructure, including REE recycling (Mancheri et al., 2019).

(c) China

China dominates the global REE market and has implemented export quotas, production caps, and environmental regulations to control the sector. Since 2010, China has emphasized **green mining**, technological innovation, and vertical integration to improve efficiency and reduce ecological damage (Wübbeke, 2013).

Policies such as the **Rare Earth Industry Development Plan (2016–2020)** promoted consolidation of mining enterprises and cleaner production processes. However, concerns remain over illegal mining and enforcement inconsistencies.

(d) India and Other Emerging Economies

India holds moderate REE reserves (notably monazite-rich beach sands) and has established initiatives like the **Rare Earth Mission** under the Department of Atomic Energy. The country's **National Electric Mobility Mission Plan (NEMMP)** supports battery and motor recycling infrastructure development (Balaram, 2019).

The KABIL initiative: Khanij Bidesh India Limited (KABIL) is a joint venture company established by three Indian public sector enterprises, namely, National Aluminium Company Ltd., Hindustan Copper Ltd., and Mineral Exploration and Consultancy Ltd. The main objective of this venture is to identify, explore, acquire, and develop mineral assets, particularly lithium and cobalt, in other countries to meet India's domestic needs. This endeavour seeks to diminish India's reliance on imports and strengthen its self-reliance in the critical minerals sector (Press Information Bureau, 2024)

Other nations, such as Australia, Canada, and Japan, are investing in REE recycling R&D and forming alliances to reduce dependence on Chinese supplies (Golev et al., 2014).

7.2 Role of Public-Private Partnerships

Government policy alone cannot address the complex challenges of REE recovery. Collaborative efforts between public agencies, private companies, and research institutions are vital.

- **Example 1:** The **REE4EU Project** (EU Horizon, 2020) demonstrated magnet-to-magnet recycling using hydrogen decrepitation, integrating industry partners like Hitachi and Siemens.

- **Example 2:** The **USA–Australia Critical Minerals Working Group** promotes joint investments in REE supply chain resilience and technology sharing (CSIRO, 2021).

Public-private partnerships (PPPs) encourage innovation, scale up pilot projects, and reduce investment risks. Effective PPPs also help bridge the gap between research outputs and market adoption.

7.3 Regulatory Challenges and Policy Gaps

Despite growing global efforts, several policy challenges hinder large-scale REE recovery:

- **Lack of Harmonized Standards:** No universal standards exist for REE recycling processes, product labeling, or environmental compliance, complicating cross-border trade and cooperation (OECD, 2019).
- **Inefficient E-Waste Collection:** Many countries lack formal e-waste recycling systems, especially in the Global South. Informal sectors dominate, often using unsafe practices that cause environmental harm and waste REEs (Balde et al., 2017).
- **Insufficient Economic Incentives:** Recycling REEs is still less economically attractive than primary mining in many regions. Subsidies, tax incentives, and green procurement policies are needed to stimulate investment.
- **Limited Data Availability:** Accurate data on REE flows, product content, and recycling rates are essential for effective policymaking but are often unavailable or unreliable (UNEP, 2013).

7.4 Future Policy Directions

To build a globally coherent REE recovery ecosystem, the following policy measures are recommended:

- **Standardization and Transparency:** Establish international benchmarks for REE recycling technologies and traceability systems.
- **Circular Economy Legislation:** Mandate product design for recyclability and ensure producer responsibility for take-back and recycling.
- **International Resource Alliances:** Strengthen multilateral cooperation for REE exploration, processing, and technology exchange.
- **Funding and Infrastructure Development:** Increase public and private investment in R&D, pilot facilities, and recycling networks.
- **Skills and Capacity Building:** Train a new generation of engineers, policy experts, and technicians in sustainable REE management.

8. Challenges, Barriers, and Research Opportunities

Regardless of the increasing concern for sustainable methods of recovery of REEs, the implementation process is constrained by a variety of hindrances posed by economic, regulatory, technological and social barriers. In order to build a secure, efficient, and environmentally sound REE supply chain, addressing these challenges and barriers is imperative. This section discusses the major obstacles and throws light upon the future research and innovation.

8.1 Technical and Operational Challenges

8.1.1 Complex Material Composition

The composition and nature of occurrence of REEs in the primary or secondary source, make their isolation and recovery complicated. The difficulties in separation and recovery may be attributed to two factors. Firstly, the low concentrations of REEs embedded within complex product matrices, as inside magnets, phosphors, and batteries. Secondly, there exists good similarity in the chemistry of these elements owing to difficulty in their individual extraction (Binnemans et al., 2013).

8.1.2 Lack of Scalable Recycling Technologies

The techniques sustainable for extraction of REEs being practised nowadays, such as *bioleaching*, *solvent extraction* with *DESs*, and *hydrogen decrepitation,* are only confined to laboratory or pilot scale (Abaka-Wood et al., 2021). These methods can be expanded for commercial operations only by incorporating process optimization, cost reduction, and robust equipment design.

8.1.3 Quality and Purity Requirements

Areas of high-performance applications, such as electronics and defence, demand extremely pure REEs. Materials obtained from recycling often require subsequent purification stages, thereby increasing energy consumption and cost, thus reducing competitiveness with virgin materials (Golev et al., 2014).

8.2 Economic and Market Barriers

8.2.1 High Recovery Costs

As compared to primary mining, the recycling process of REEs is currently more expensive. Major verticals are labour, energy, and processing costs (Jordens et al., 2013). The recycling operations are usually carried out through a decentralized mechanism incurring more per unit cost due to lack of economy of scale.

8.2.2 Price Volatility

Uncertainties in the supply chain due to evolving geopolitical developments have resulted in frequent price fluctuations in the REE market. Due to this, investment in recovery infrastructure often becomes a bit risky for private enterprises (Mancheri et al., 2019).

8.2.3 Low Collection Rates

For urban mining to be viable, a consistent feedstock of REE-rich waste is required. However, **collection rates for e-waste** remain low globally—only 17.4% was formally collected in 2019 (Forti et al., 2020), limiting resource availability for recyclers.

8.3 Regulatory and Institutional Constraints

8.3.1 Weak Enforcement

In many countries, especially in the Global South, regulations on e-waste handling, pollution control, and occupational safety are weak or poorly enforced. Informal recycling often operates outside legal frameworks, with **minimal environmental oversight** (Leung et al., 2021).

8.3.2 Fragmented Policy Landscape

Global REE policies lack coordination, and few international treaties address resource recovery explicitly. This **regulatory fragmentation** hinders the establishment of shared standards, traceability systems, and best practices (OECD, 2019).

8.3.3 Intellectual Property Barriers

Some efficient recycling technologies are **patent-protected**, restricting their wider adoption in developing regions. Open innovation and technology licensing will be essential to bridge this gap.

8.4 Social and Behavioural Barriers

8.4.1 Consumer Awareness

A significant portion of REE-containing devices ends up in households or landfills because consumers are unaware of their value or lack access to recycling facilities (Balde et al., 2017). Effective public education and return incentives are needed.

8.4.2 Labour and Health Risks

Informal workers in developing countries extract REEs under unsafe conditions, using **hazardous chemicals and manual tools**. Improving working conditions and formalizing the informal sector are critical for equitable and sustainable REE recovery (Ali, 2014).

8.5 Research and Innovation Opportunities

Despite these barriers, several promising research directions can enhance REE sustainability:

Table 3 Summary of Research and Innovation Opportunities in REE.

Focus Area	*Key Challenges/Barriers*	*Research/ Innovation Opportunities*
Technological Limitations	• Inefficient recovery rates • Poor selectivity • Lack of scalable eco-friendly methods	Development of bioleaching, DES, and AI-based recycling systems Process intensification
Economic and Market Barriers	High cost of recycling vs. mining Price volatility of REEs Lack of business incentives	Life-cycle cost-benefit analyses Green tax incentives Circular business models
Policy and Regulatory Gaps	Weak enforcement of e-waste laws Lack of harmonized global standards Informal sector dominance	International regulatory alignment EPR expansion Formalization of recycling networks
Research and Knowledge Gaps	Limited data on REE flow • Few pilot-scale technologies • Fragmented interdisciplinary work	System-level modeling of RE E flows • Cross-disciplinary collaboration • Open innovation platforms

- **Green Solvents:** Development of non-toxic, biodegradable leaching agents for high-selectivity REE extraction (Sheng et al., 2020).
- **Advanced Membrane Technologies:** Use of molecularly imprinted polymers and nanofiltration membranes for precise REE separation.
- **Rare Earth Based Hybrid Composites:** Development of advanced hybrid composites with REEs as reinforcement materials for advanced applications like aerospace, automobile and industrial purposes.
- **AI and Machine Learning:** Optimization of recycling processes through predictive models and smart sorting systems (Gautam et al., 2022).
- **Design for Recycling:** Embedding circularity into product design to ease disassembly and material recovery.
- **Techno-Economic and LCA Models:** Holistic evaluation of environmental impact, cost-efficiency, and resource circularity for new recovery techniques.

These research priorities can guide academia, industry, and governments toward a more resilient and sustainable REE ecosystem. The Research and Innovation Opportunities are summarized in Table 3.

9. Conclusion

Today, the sustainable supply of REEs finds a pivotal role in driving the current global shift to a carbon-neutral, digitally sound and connected, and resource-efficient economy. These materials, although present in small amounts, play a vital role in critical sectors such as renewable energy, electric mobility, electronics, defence, and medical imaging. However, the current dependence on primary mining of REEs presents a set of challenges, including environmental degradation, geopolitical fluctuations, and resource depletion.

In view of the limitations of conventional REE supply chains, adapting circular economy models and sustainable recovery technologies is imperative in order to ensure long-term material security. Exploration of secondary sources like E-waste, industry residues, end-of-life green technologies, etc., for REE recycling is not only viable but also essential for environmental and economic sustainability.

The emerging green extraction techniques like *hydrometallurgy*, *pyrometallurgy*, *biometallurgy*, and *AI-enabled sorting systems* are very promising for REE recovery yet these are subject to constraints in terms of scalability, economic feasibility, and purity control. Moreover, the enabling frameworks for circular economy have to be framed promptly. Models like *PaaS, Eco-design*, *urban mining*, and *EPR* are the vital components of a circular system for REEs.

With regard to policy perspective, various countries are taking strategic steps toward critical material security. The European Union, the United States, China, India, and other nations have initiated national programmes, trade collaborations, and PPPs to develop domestic recycling capacity and minimize the dependence on import. However, disintegrated regulation, lack of infrastructure, and low public awareness has continued to retard the progress.

Multidisciplinary research and coordinated international action are crucial in order to bridge these gaps. The way forward to stabilizing the REE supply chain ensuring the resource security, should consider following aspects:

- Increased investment in research development for scalable green technologies;
- Development of standardized recycling protocols;
- Strengthening regulatory frameworks and enforcement mechanisms;
- Encouraging open innovation, collaborations and knowledge-sharing;
- Enhancing consumer awareness and reverse logistics networks.

Finally, the need for sustainable recovery and reuse of REEs is inevitable as it not just forms the technical or economic solution, but also plays a pivotal role in bringing about the transition toward a clean, resilient, responsible, and sustainable future.

With the drive towards a future framed by clean technologies and circular resource flows, REEs continue to act as both a challenge and an opportunity. Harnessing their potential, accompanied with efforts to mitigate their impact, requires visionary policies, technological innovation, and global collaboration.

References

Alves Dias, P., Bobba, S., Carrara, S. and Plazzotta, B. (2020). *The Role of Rare Earth Elements in Wind Energy and Electric Mobility. JRC Science for Policy Report*. European Union.

Abaka-Wood, G.B., Ofori-Sarpong, G. and Dodoo-Arhin, D. (2021). Recent developments in hydrometallurgical recovery of rare earth elements from secondary resources. *Minerals Engineering*, *160*, 106639. https://doi.org/10.1016/j.mineng.2020.106639.

Abrahami, S.T., van den Boogaart, K.G. and Binnemans, K. (2015). Selective separation of rare earths using ionic liquids. *Green Chemistry*, 17(8), 4382–4392. https://doi.org/10.1039/C5GC00855A.
Ali, S.H. (2014). Social and environmental impact of the rare earth industries. *Resources*, 3(1), 123–134. https://doi.org/10.3390/resources3010123.
Alonso, E., Sherman, A.M., Wallington, T.J., Everson, M.P., Field, F.R., Roth, R. and Kirchain, R.E. (2012). Evaluating rare earth element availability: A case with revolutionary demand from clean technologies. *Environmental Science and Technology*, 46(6), 3406–3414. https://doi.org/10.1021/es203518d.
Balde, C.P., Forti, V., Gray, V., Kuehr, R. and Stegmann, P. (2017). *The Global E-waste Monitor 2017*. United Nations University. https://ewastemonitor.info/
Balaram, V. (2019). Rare earth elements: A review of applications, occurrence, exploration, analysis, recycling, and environmental impact. *Geoscience Frontiers*, 10(4), 1285–1303. https://doi.org/10.1016/j.gsf.2018.12.005.
Binnemans, K., Jones, P.T., Blanpain, B., Van Gerven, T., Yang, Y., Walton, A. and Buchert, M. (2013). Recycling of rare earths: a critical review. *Journal of Cleaner Production*, 51, 1–22. https://doi.org/10.1016/j.jclepro.2012.12.037.
Bocken, N.M.P., Bakker, C. and Pauw, I.D. (2016). Product design and business model strategies for a circular economy. *Journal of Industrial and Production Engineering*, 33(5), 308–320. https://doi.org/10.1080/21681015.2016.1172124.
Borai, E.H., Sharshar, T., El-Ramady, H. and Eldridge, R. (2020). Recovery of rare earth elements from phosphogypsum: A review. *Journal of Environmental Management*.
Brunner, P.H. (2011). Urban mining: A contribution to reindustrializing the city. *Journal of Industrial Ecology*, 15(3), 339–341. https://doi.org/10.1111/j.1530-9290.2011.00345.x.
Castor, S.B. and Hedrick, J. B. (2006). Rare earth elements. In J. E. Kogel, N. C. Trivedi, & J. M. Barker (Eds.), Industrial minerals & rocks: Commodities, markets, and uses (7th ed., pp. 769–792). Society for Mining, Metallurgy, and Exploration, Inc.
CSIRO. (2021, February 25). Rare earths and critical minerals provide significant opportunities for Australia.
DOE (U.S. Department of Energy). (2022). Critical Materials Institute: Annual Report. https://cmi.ameslab.gov.
Ellen MacArthur Foundation. (2015). *Towards the Circular Economy: Economic and Business Rationale for an Accelerated Transition*. https://ellenmacarthurfoundation.org.
European Commission. (2023). *Critical Raw Materials Act – Factsheet*. https://ec.europa.eu.
Forti, V., Baldé, C.P., Kuehr, R. and Bel, G. (2020). *The Global E-waste Monitor 2020*. United Nations University. https://ewastemonitor.info.
Gautam, R., Kim, J., Kim, J.W. and Kim, H.S. (2022). Machine learning-based optimization for rare earth element recovery from e-waste. *Journal of Cleaner Production*, 357, 131803. https://doi.org/10.1016/j.jclepro.2022.131803.
Geissdoerfer, M., Savaget, P., Bocken, N.M.P. and Hultink, E.J. (2017). The circular economy: A new sustainability paradigm? *Journal of Cleaner Production*, 143, 757–768. https://doi.org/10.1016/j.jclepro.2016.12.048.
Golev, A., Scott, M., Erskine, P.D., Ali, S.H. and Ballantyne, G.R. (2014). Rare earths supply chains: Current status, constraints, and opportunities. *Resources Policy*, 41, 52–59. https://doi.org/10.1016/j.resourpol.2014.03.004.
Gupta, C.K. and Krishnamurthy, N. (2005). *Extractive Metallurgy of Rare Earths*. CRC Press.
Humphries, M. (2013, December 16). Rare Earth Elements: The Global Supply Chain. Congressional Research Service.
International Energy Agency. (2024). Mineral requirements for clean energy transitions: The role of critical minerals in clean energy transitions (IEA Report).
Jordens, A., Cheng, Y.P. and Waters, K.E. (2013). A review of the beneficiation of rare earth element bearing minerals. *Minerals Engineering*, 41, 97–114. https://doi.org/10.1016/j.mineng.2012.10.017.

Kumar, A, Yatika G., Chandan SM., Singh V.P. and Sharma V.K. (2022). Design and analysis of heavy vehicle medium duty transmission gearbox system. In: *Additive Manufacturing in Industry 4.0*, pp. 67–86. CRC Press.

Leung, D.Y.C., Caramanna, G. and Maroto-Valer, M.M. (2021). A critical review of rare earth resources, extraction, and utilization: Implications for clean energy and sustainability. *Journal of Cleaner Production*, 289, 125654. https://doi.org/10.1016/j.jclepro.2020.125654.

López, M., Pena-Castro, J.M. and Gallego, C. (2019). Rare earth element phytoremediation: Current advancement in exploitation of hyperaccumulator plants. *Journal of Hazardous Materials*, 380, 120869.

Mancheri, N.A., Sprecher, B., Bailey, G., Ge, J. and Tukker, A. (2019). Effect of Chinese policies on rare earth supply chain resilience. *Resources, Conservation, and Recycling*, 142, 101–112. https://doi.org/10.1016/j.resconrec.2018.11.017.

Massari, S. and Ruberti, M. (2013). Rare earth elements as critical raw materials: Focus on international markets and future strategies. *Resources Policy*, 38(1), 36–43.

Nancharaiah, Y.V. and Mohan, S.V. (2022). Bio-recovery of rare earth elements by acidophilic microorganisms: *A review. Bioresource Technology*, 351.

Ochsenkühn-Petropoulou, M., Tüfekçi, N., & Dietz, P. (2002). Pilot-plant investigation of the leaching process for the recovery of scandium and lanthanides from red mud. *Industrial & Engineering Chemistry Research*, 41(23), 5903–5907.

OECD. (2019). *Global Material Resources Outlook to 2060*. Organisation for Economic Co-operation and Development. https://www.oecd.org/environment/global-material-resources-outlook-2060-9789264307452-en.htm.

Patel, Dinesh Kumar, Sachin Kumar, Vipin Kumar Sharma, Hari Om Sharma, & Arbind Prasad. (2024). Alternative Source Systems of In-Vehicle Electricity Production. In: *Sustainable Mobility: Policies, Challenges, and Advancements*, pp. 61–78. Wiley-Scrivener.

Patel, D.K. and Sharma V.K. (2025). Illuminating Innovations: A Comprehensive Review of Rare Earth-Doped Phosphors and Their Applications in LEDs. *Transactions of the Indian Institute of Metals*, 78(7), 165.

Patel, D.K., Sharma V.K., Kumar V., Kumar P. and Kumar D. (2025). Enhancing hydrophobicity and anti-corrosion properties of Al-6061Aluminum hybrid composites through trace additions of rare earth oxides. In: *Proceedings of the Institution of Mechanical Engineers, Part E: Journal of Process Mechanical Engineering*, 09544089251317381.

Press Information Bureau. (2025, July 23). Parliament Question: Rare Earth Minerals [Press release]. https://www.pib.gov.in/PressReleasePage.aspx?PRID=2147282.

Press Information Bureau. (2024, July 31). KABIL is exploring opportunities for acquisition of overseas critical minerals assets in Argentina, Australia and Chile. [Press release]. https://www.pib.gov.in/PressReleaseIframePage.aspx?PRID=2039606.

REE4EU. (2020). Magnet-to-Magnet Recycling: Hydrogen Decrepitation Process. [Project deliverable]. EU Horizon 2020.

Rim, K.-T., Kim, J.-S. and Yang, J.-S. (2013). Toxicological evaluations of rare earths and their health impacts to workers: A literature review. *Safety and Health at Work*, 4(1), 12–26.

Schreiber, A., Kynicky, J., Volek, M. and Evans, K. (2016). Rare-earth element recovery from end-of-life permanent magnet machines: A sustainable supply chain perspective. *Journal of Cleaner Production*, 137, 1251–1263.

Sehgal, A., Sharma V.K., Gautam P.S. and Partap A.. (2025). Forging the Future: Exploring Aluminium AA-6061 *Composites through Stir Casting for Advanced Applications and Future Innovations*, pp. 806–826. ResearchGate.

Sharma, V.K., Kumar V. and Joshi R.S. (2019). Effect of RE addition on wear behavior of an Al-6061 based hybrid composite. *Wear*, 426, 961–974.

Sharma, V.K., Kumar V. and Joshi R.S. (2019). Investigation of rare earth particulate on tribological and mechanical properties of Al-6061 alloy composites for aerospace application. *Journal of Materials Research and Technology*, 8(4), 3504–3516.

Sharma, V.K. and Kumar V. (2019). Development of rare-earth oxides based hybrid AMCs reinforced with SiC/Al2O3: Mechanical and metallurgical characterization. *Journal of Materials Research and Technology*, 8(2), 1971–1981.

Sharma, VK., Kumar V. and Joshi R.S. (2019). Experimental investigation on effect of RE oxides addition on tribological and mechanical properties of Al-6063 based hybrid composites. *Materials Research Express*, 6(8), 0865d7.

Sharma, V.K., Kumar V. and Joshi R.S. (2020). Parametric study of aluminium-rare earth based composites with improved hydrophobicity using response surface method. *Journal of Materials Research and Technology*, 9(3), 4919–4932.

Sharma, V.K., Aggarwal D., Kumar V. and Joshi R.S. (2021). Influence of rare earth particulate on the mechanical and tribological properties of Al-6063/SiC hybrid composites. *Particulate Science and Technology*, 39(8), 928–943.

Sharma, V.K., Kumar P., Kumar D., Chalisgaonkar R. and Akhai S. (2024). Characterization of Al 6061/Al2O3/SiC composites with cerium oxide: Corrosion analysis and microstructural insights. *Composites Theory and Practice*, 24.

Sheng, J., Liu, J. and Sun, Z. (2020). Application of deep eutectic solvents for selective separation of rare earths: Progress and perspectives. *Separation and Purification Technology*, 248, 117034. https://doi.org/10.1016/j.seppur.2020.117034.

Srikant, Sharma V.K., Kumar V., Joshi R.S., Jain S., and Kumar P. (2021). A review of recent research on rare earth particulate composite materials and structures with their applications. *Transactions of the Indian Institute of Metals*, 74(11) (2021): 2569–2581.

Stahel, W.R. (2016). The circular economy. *Nature*, 531(7595), 435–438. https://doi.org/10.1038/531435a.

Taggart, R.K., Bragg, L.M. and Reddy, M.S. (2018). Effects of roasting additives and leaching parameters on rare earth extraction from coal fly ash. *Fuel Processing Technology*, 176, 128–137.

Tan, M., Bie, M. and Zhang, L. (2021). Artificial intelligence and robotics in recycling: disassembling and recovering critical materials from e-waste. Resources, *Conservation & Recycling*, 169.

UNEP (2013). Addressing the environmental dimensions of critical minerals: Environmental risks across the life cycle of minerals. United Nations Environment Programme.

United Nations Institute for Training and Research & International Telecommunication Union. (2024). The Global E-Waste Monitor 2024 (4th ed.).

U.S. Department of Energy. (2011). Critical Materials Strategy. U.S. DOE.

U.S. Geological Survey (USGS). (2023). Mineral Commodity Summaries: Rare Earths.

Walton, A., Williams, A.J., Mann, V.S.J., Sheridan, R., Speight, J.D. and Harris, I.R. (2015). The use of hydrogen to separate and recycle neodymium–iron–boron-type magnets from end-of-life products. *Journal of Cleaner Production*, 104, 236–241. https://doi.org/10.1016/j.jclepro.2015.04.108.

Wang, H., Lin, Q., Xie, Y. and Liu, H. (2021). Biosorption of rare earth elements by microorganisms: A review. *Chemosphere*, 265, 129037. https://doi.org/10.1016/j.chemosphere.2020.129037.

Widmer, R., Seiler, A. and Wäger, P. (2015). Secondary resources review: Rare earth elements. Paul Scherrer Institut (PSI). (PSI report).

Wübbeke, J. (2013). Rare earth elements in China: Policies and narratives of reinventing an industry. *Resources Policy*, 38(3), 384–394. https://doi.org/10.1016/j.resourpol.2013.05.004.

Xie, F., Zhang, T.A., Dreisinger, D. and Doyle, F. (2021). Recovery of rare earths from ion-adsorption clays: Review and future perspectives. *Hydrometallurgy*, 203, 105688. https://doi.org/10.1016/j.hydromet.2021.105688.

Yang, Y., Walton, A., Sheridan, R., Güth, K., Gauß, R., Gutfleisch, O. and Buchert, M. (2022). REE recovery from end-of-life NdFeB magnets: A review of technology, sustainability, and economics. *Journal of Sustainable Metallurgy*, 8(2), 285–312. https://doi.org/10.1007/s40831-021-00471-3.

Zeng, X., Li, J. and Singh, N. (2016). Recycling of rare earths from spent fluorescent lamps: A review. *Critical Reviews in Environmental Science and Technology*, 46(3), 198–239. https://doi.org/10.1080/10643389.2015.1061867.

Zhang, Y., Liu, J., Wang, L. and Sun, Z. (2022). Advances in deep eutectic solvents for the separation of rare earths. *Separation and Purification Technology*, 290, 120912. https://doi.org/10.1016/j.seppur.2022.120912.

2 AI-Driven Innovations in Rare Earth Element Processing and Separation Techniques: Towards Precision and Sustainability

Sharad Kumar[1], Vikas Sharma[2*], Tarun Kumar Vashishth[3], Sanjukta Vidyant[4], Avitaj Kaushal[5] and Shahanawaj Ahamad[6]

1. Introduction

Scandium and yttrium, along with the 15 lanthanides, form the Rare Earth Elements (REEs) critical to the advancement of contemporary technologies. Magnets used in electric vehicles and wind turbines, display screens and lights, batteries, and even auxiliary components in medical imaging and defense systems require permanent magnets to function (Sharma et al., 2021, Sharma et al., 2024). Additionally, these with phosphors and display screens, along with catalysts in the automotive and petroleum industries, and medical imaging have critical purposes in modern life. The ongoing shift towards clean energy, smart technologies, and digital infrastructure has a global increase in demand for REEs. However, there are challenges accompanying the need such as supply chains and environmental sustainability issues fully dedicated to a few countries holding the concentration of extracting and processing the REEs (Delogu et al., 2023).

1.1 Importance of Rare Earth Elements (REEs) in Modern Technology

The physical attributes and chemical properties of REEs like their magnetism, electrical conductivity, and luminescence make them critical to modern technologies. These elements form the backbone of many advanced applications in renewable energy, electronics, healthcare, automotive, aerospace, and defense

[1] School of Engineering and Technology Shri Venkateshwara University, Gajraula, Amroha, U.P., India.
[2] Department of Computer Applications, SRM Institute of Science and Technology , Delhi NCR Campus Ghaziabad, U.P., India.
[3] School of Computer Science and Applications, IIMT University, Meerut, U.P., India.
[4,5] Department of Biotechnology, Shobhit University, Meerut, U.P., India.
[6] College of Computer Science and Engineering, University of Hail, Saudi Arabia.
* Corresponding author: vicky.c610@gmail.com

DOI: 10.1201/9781003683254-2

industries (Srikant et al., 2021; Patel et al., 2024; Patel et al., 2025). For example, high-performance neodymium and dysprosium permanent magnets are used in electric vehicle and wind turbine motors as well as in robotics. Europium and terbium are active components of LEDs, LCDs, and even plasma screens used in modern energy efficient lighting and display technology (Patel et al., 2025). In healthcare, gadolinium is used widely as a contrast agent in MRI scans, cerium and lanthanum are used in catalytic converters alongside their use in optical lenses. Moreover, REE's ability to enhance smartphone, laptop, and sensor functionalities allow for the miniaturization of these devices. The REEs are becoming increasingly important for other technologies integrated with infrastructure aimed at providing clean energy sources accessible around the world. Their position in the growing technologies like electric mobility, grid-scale batteries, advanced defense systems, and quantum computing demonstrates their importance both as economic and geopolitical resources. The limited availability and geopolitical concentration of the materials has created the necessity for developing sustainable REE processing technologies (Vashishth et al., 2024).

1.2 Need for Sustainable and Intelligent Processing Approaches

Now, the readily available modern technologies and clean energy in conjunction with the diverse strategic fields that incorporate REEs highlight the need to reevaluate the traditional methodologies used for processing the mineral constituents. The extraction and separation techniques that utilize hydrometallurgical and pyrometallurgical processes are environmentally damaging and extremely resource intensive. So is the requirement of extracting dangerous chemicals, such as strong acids, organic solvents, and the emission of radioactive and toxic waste alongside water and energy in large quantities. In addition, the chemical analogy of REEs enhances the technical and economic inefficiency of selective separation leading to sub-optimal recovery rates and higher operational costs. The situation is exacerbated through the uneven distribution of available REE reserves within the globe causing vulnerabilities to supply chains and geopolitical conflicts (Adisyahputra et al., 2024). Because of these problems, it is clear that new intelligent and sustainable processing approaches need to be created to address economic and environmental issues simultaneously. Artificial Intelligence (AI) acts as a solution due to its data processing abilities, modeling intricate chemical systems, and real-time parameter optimization. Incorporating AI techniques into REE processing workflows could improve selectivity, recovery efficiency, energy consumption, and waste generation. Predictive maintenance, automated quality control, and adaptability to different feedstock compositions can be enabled by intelligent systems, improving the resilience and cost-effectiveness of the entire processing chain. In light of the above, implementing AI is a critical step towards advanced circularity and lowering environmental footprint of rare earth supply chains in alignment with global sustainability efforts (Chittalury, 2025).

1.3 Role of Artificial Intelligence in Metallurgy

AI stands out as a new complex technology in the field of metallurgy as it reinvents the methods of material inception, extraction, processing, and optimization. When it comes to processing REEs—where chemistry, operation, and ecology bottlenecks productivity—AI offers unprecedented precision, flexibility, and sustainability. With machine learning (ML), deep learning, and data analytics, AI systems can scrutinize voluminous datasets from mines, labs, and real-time sensors to track dependencies and relationships that are not possible for a human to fathom. This enhances decision-making at every level of the metallurgical cycle including ore characterization, beneficiation, separation, and purification. Predictive modeling is one of AI's hallmark contributions where algorithms can estimate how a certain ore body will behave with a predetermined set of treatment options, allowing setting of process parameters for maximum output and minimum waste. In separation processes—most difficult for REEs because of their low, almost identical, solubility constants—AI can optimize selectivity and yield by adjusting pH, temperature, and solvent concentration. AI-enabled real-time monitoring systems can ensure product and process quality consistency by automatically adjusting control measures to predefined levels while maintaining their safety through process deviation detection using spectroscopy, imaging, and sensor fusion technologies (Rojas et al., 2025). Moreover, AI helps create digital twins, which are virtual simulations of physical processes that allow for operational scenario testing without real-world disruption. These environments allow for outcome prediction, strategy testing for new separations, and risk assessments that are cost-effective and non-disruptive. The application of reinforcement learning—where systems learn optimal actions through interaction for autonomous decision-making—has grown to include the dynamic control of complex extraction systems that must adapt to changing operational conditions. With these capabilities, AI optimization not only decreases energy and reagent expenditures, but also increases the level of automation, reduces human engagement in dangerous conditions, and mitigates human exposure to risk. From the perspective of sustainable metallurgy, AI enables intelligent waste retrieval, secondary resource recovery, and more efficient recycling of REE-containing products, thereby supporting circular economy principles. The integration of AI into metallurgy signifies a transformation toward more advanced, automated, responsive, and efficient technologies that integrate concern for the environment, resource conservation, supply chain robustness, and energy management (Bashynska & Prokopenko, 2024; Patel et al., 2024).

2. Overview of Rare Earth Element Processing Techniques

The energy costs involved with sophisticated computation and multi-stage separation methods using solvents are extreme. The complex associations between REEs with other minerals and the geochemical similarities among them result in intricate multi-stage processes that are both technically difficult and labor-intensive. Even though the Earth's crust contains REEs in relatively abundant amounts, their economical concentration is seldom feasible, making extraction difficult and resource-intensive. The processing starts with the mining and beneficiation phase, wherein REE-bearing minerals like bastnäsite, monazite, and xenotime, are crushed, ground, concentrated, and extracted using flotation and gravity separation techniques. This is succeeded by hydrometallurgical and pyrometallurgical processes aimed at dismantling the mineral matrix to recover individual REEs, followed by chemical extraction processes.

2.1 Mining and Ore Beneficiation

The first step in the production of REEs is the mining and beneficiation of REE-bearing ores since it can determine the productivity and ecological consequences of the REE production cycle. REEs are usually present in extremely low amounts in fusion rocks and are almost always associated with bastnaesite, monazite, xenotime, allanite, or some other minerals. These minerals often occur in solid rock as well as placer deposits, requiring energy-consuming extraction methods like open-pit or underground mining, depending on the depth and nature of the deposit. There is some raw ore lying about, but its mined minerals have to go through a multitude of physical and chemical processes to concentrate and isolate the populous minerals containing REEs from the barren materials surrounding them. Crushing, followed by grinding, enhances the ore's grade and reduces its volume in relation to other processing, as it usually does with the ore's aggregate. After the comminution processes, selective separation is carried out employing methods such as gravity concentration, magnetic separation, electrostatic separation, and froth flotation to extract the REE minerals. For instance, magnetic separation is applicable to bastnäsite and monazite because of their paramagnetic characteristics. Others, like flotation, are well established in separating REE minerals from the waste material based on differences of surface chemistry. The success of these approaches' branches of different ore types, particle size distribution, and mineralogical composition. Sustainable beneficiation processes are crucial in increasing the output of REEs from the ore while minimizing the energy and chemicals used in the following processing steps. Additionally, the production and control of tailings and radioactive waste—in particular from monazite ores which typically have thorium—pose serious environmental and administrative issues. Due to this, the use of smart systems with sensor-based sorting technologies at the mining and beneficiation stages is increasing. With these advancements, recognition, efficient separation of minerals, and resource waste in separation processes are done in real-time. Generally, all the methods of mining and beneficiation of the REEs contribute towards the economically sustainable recovery of the materials which form the backbone of modern-day technology (Liang et al., 2024).

2.2 Hydrometallurgical Techniques

Among several techniques used to process REEs, hydrometallurgy takes the lead. This method utilizes aqueous chemistry to isolate and refine REEs from minerals, recovering them after they are incorporated into the ores. This is done after mining and beneficiation when concentrated ores are subjected to various mechanical processes to break them down to simpler parts and individual elements. It is preferred over thermal methods due to selectivity and efficiency. A significant part of the refining is done by important hydrometallurgical processes such as leaching, solvent extraction, ion exchange, and precipitation as they are all important to some extent in the isolation and refining of REEs. Alkaline or acidic solutions are applied to REE concentrated minerals in the first stage called leaching, to convert certain ions of rare earth elements into liquid form. This step requires significant amounts of energy; therefore, the leaching agents are typically sulfuric acid, hydrochloric acid, and sodium hydroxide for most ores. Choosing the correct agent depends on the environmental factors and minerals present in the ore. This phase usually requires temperature control and agitation to balance the dissolution of valuable materials and impurities which are best avoided. After the leaching stage, REE separation is mostly done through solvent extraction where the selective affinities of particular REEs to specific organic solvents are used. Selective separation can be accomplished through the small differences in the ionic radii and complexing properties of the REEs. Organic solvent containing extractants, such as tributyl phosphate (TBP) or organophosphorus compounds, attach 'target' ions, which are stripped into a purified aqueous phase. Ion exchange complements solvent extraction because of their ability to further clean the precious REEs using resins that claim rare earth ions from solutions. Recovery is finalized through the precipitation and crystallization steps where the REE ions are transformed into solid compounds—usually oxalates or carbonates—which can be calcined to yield rare earth oxides, the commercially preferred form for secondary services. In comparison to routes that employ high temperatures, hydrometallurgical methods are much more appealing due to their lower environmental impact, lower temperature, and ease in control and adaptability. However, challenges remain in handling large volumes of acidic or alkaline effluents, managing radioactive by-products, and achieving high selectivity for closely related REE species (Vashishth et al., 2024). The most recent efforts combine AI with sensor technologies to optimize not only leaching but also kinetic parameters: the composition of the solvent and separations done all in real-time. This modern refinement results in lower consumption of chemicals, energy, and waste. In general, the use of hydrometallurgical techniques is required for the sustainable, accurate, and economical extraction and separation of REEs as noted by (Vashishth et al., 2024).

3. Pyrometallurgical Techniques

These methods are particularly important in the processing of minerals that are resistant to chemical leaching or need to undergo some transformation into more approachable varieties for further treatment. They are the high-temperature processing techniques which involve the use of heat in changing the chemical and physical structures of ores or concentrates to extract and refine REEs. Common pyrometallurgical processes are roasting, calcination, smelting, and molten salt electrolysis. All of them assist other pyrometallurgical processes such as dismantling mineral matrices, removing non-desired materials, and isolating key elements. Both roasting and calcination processes transform carbonate, sulphate, or other compounds into oxides through complete decomposition when oxygen or some other gas is present. This step is needed in processing ores like bastnäsite and monazite because it makes them more amenable to hydrometallurgical subsequent steps. For instance, roasting monazite at high temperatures oxidizes phosphorus and thorium contaminants, thus enabling safer further processing downstream. Smelting processes include the processes of melting the ore concentrate with fluxes and reductants at very high temperatures where metals separate from gangue by difference in melting point, density, and some degree of chemical affinity. Smelting has the ability of forming intermediate products like REE-rich alloys or slags that are later purified, thus refined. Dubbed as energy-sapping, smelting is nonetheless useful in concentrating REEs and removing impurities in a single step (Ahmed et al., 2025). An innovative approach to extracting REEs is through molten salt electrolysis, which comprises a pyrometallurgic step where an REE oxide is dissolved in molten salt and subsequently electrolytically reduced to the pure metal. This technique allows for the direct production of REE metals that are essential in advanced technologies, such as permanent magnets and other electronic parts. On the other hand, a specific temperature, composition of the electrolyte, and materials of the electrode interfaces need to be strictly controlled in order to maximize purity and yield. On the other hand, most studies combine the pyrometallurgical approach with hydrometallurgical techniques to maximize recovery. This so-called 'integrated' technique is known to yield, however, comes at the cost of higher energy consumption and greenhouse gases. The extraction, as well as the disposal of radioactive by-products, especially those from monazite, present further environmental and safety issues. More modern focuses dealt with the optimization of energy consumption through AI and sensor networks to monitor furnace processes towards improved product quality, further enabling real-time temperature adjustments and process alterations, which are traditionally more precise through AI. This inherently renders the environment more sustainable while modernizing the extraction of REEs (Singh et al., 2025).

4. Environmental Impacts and Limitations of Current Techniques

While REEs are required for modern technologies, their extraction and exploitation is a process that poses serious disentanglement issues for the environment which has started to become a major issue worldwide. Both traditional mining and processing techniques using the hydrometallurgy and pyrometallurgy approach is the problem with the expenditure of enormous amounts of hazardous materials. These processes consume enormous amounts of energy and generate poisonous waste which further endangers the ecological and health balance of the locality. Acid leaching processes, for example, release not only heavy metals but also radioactive elements such as thorium and uranium and much more, which is harmful to the soil and water bodies if not taken care of. Similarly, pyrometallurgical processes are not cleaner. They discharge greenhouse gases and particulate matter which increase air pollution and further amplify global warming. Beyond pollution, the inground space occupied by mining works where there are open pits, waste rock dumps, and tailings ponds is an eye sore that further paves way for the loss of land, soil erosion, and biodiversity. Contained, the tailings produced during the beneficiation and chemical treatment also contain radioactive and poisonous materials which presents potential negative consequences from an ecological standpoint should the containment structures collapse (Yin et al., 2025). Moreover, the consumption of water in these operations is significant, putting pressure on local water supplies, particularly in dry areas. From a technological perspective, current REE processing methods using hydrometallurgy are deficient in efficiency, selectivity, and scope. The chemical similarity of REEs makes their separation difficult, often requiring several energetically costly stages that increase operational expenditure and harm the environment. Recovery is particularly inefficient for low-grade ores or complex mixed deposits, resulting in considerable material wastage. In addition, the management and disposal of radioactive waste, particularly from monazitic ores, has near-perpetual regulatory and safety implications (Kesar & Joseph, 2025).

5. AI Technologies Empowering REE Processing

AI is one of the most powerful emerging technologies that is taking the world by a storm, especially with REE processing. It simplifies a lot of processes and tasks when it comes to dealing with inefficiencies and complexities present in the old methods. Now, stakeholders can utilize advanced data analytics, ML algorithms, and even intelligent automation to improve accuracy, achieve better resource optimization, and increase efficiency when it comes to using natural AI technologies, which reduce negative environmental effects. AI's ability to analyze data is truly prolific; not only can it do so in real-time, but it can also manage heterogeneous datasets.

5.1 Machine Learning Models for Predictive Analytics

ML models are now used in the forecasting of REE processing to aid decision-making for greater performance and reliability in operations (Sharma et al., 2022; Sharma et al., 2022). With modern ML algorithms, it is programmed to detect different patterns from historical and real-time data which allows for accurate forecasting of extraction yields, reagent consumption, and process throughput. The ability operators have to foresee outcomes is advanced enabling proactive modifications of process control parameters that if unattended would cause disruption. Various supervised learning techniques that include regression models, support vector models, and neural networks are now being implemented to automate the capture of numerous nonlinear relationships between input factors and process outputs. As an illustration, an ML model can be taught to calculate the optimal temperature and acid concentration for maximum dissolution of REE while minimizing impurity extraction for leaching processes. Unsupervised learning access techniques such as clustering and dimensionality reduction are able to facilitate the classification of ore types as well as the identification of unusual process behavior that may signal malfunctioning equipment or less favorable operating conditions. The integration of data from spectroscopy, particle size analyzers, and flow meters provides constant real-time data streams to ML systems, which aids in dynamic predictive control. Adaptive process optimization is particularly fostered by feedback loop policy RL models due to the varying processes from ore heterogeneity or outside perturbations. Figure 1 illustrates a typical ML architecture used for predictive analytics, showing stages like data collection, preprocessing, model training, evaluation, and prediction. It highlights how data flows through the system to generate actionable insights.

These models, as described by Zhang et al. (2025), advance by testing numerous control strategies and assimilating which ones produce favorable impacts. Therefore, the implementation of ML-based predictive analytics optimizes the accuracy and flexibility regarding REE processing operations by decreasing the chemicals, energy, and wastes emitted. These shifts enable the extraction and separation of REEs to become more safe, environmentally friendly, and economical management systems, as the models work to reduce reactionary approaches.

5.2 Deep Learning for Mineral and Phase Identification

The application of AI, specifically deep learning, which is modeled after the neural networks of the human brain, has automated and greatly improved the precision of mineral and phase identification in the processing of REEs by analyzing intricate geological datasets. Conventional methods of identifying minerals often employ manual inspection through a microscope or simplistic spectral analysis, which are inaccurate, inefficient, and extraordinarily subjective, particularly for finely intergrown minerals or complex ore mixtures. Convolutional neural networks (CNNs), a type of deep learning model, are particularly adept at feature extraction at scale and have advanced imaging and spectral data processing, which allows for the fast classification of minerals and their phases. These models are trained on labeled datasets such as hyperspectral images, X-ray diffraction patterns, and scanning electron microscopy (SEM) images, where they learn to differentiate textural, morphological, and compositional subtleties that characterize mineral species and phases. Deep learning's automation of the identification process, as compared to traditional methods, means that human error is virtually removed, and analysis is conducted at much greater speeds, allowing for real-time processing feedback. As an illustration, exact phase recognition ensures that the pertinent feed materials are within specified quality limits before expensive downstream processing, optimizing beneficiation steps. Also, the integration of visual, spectral,

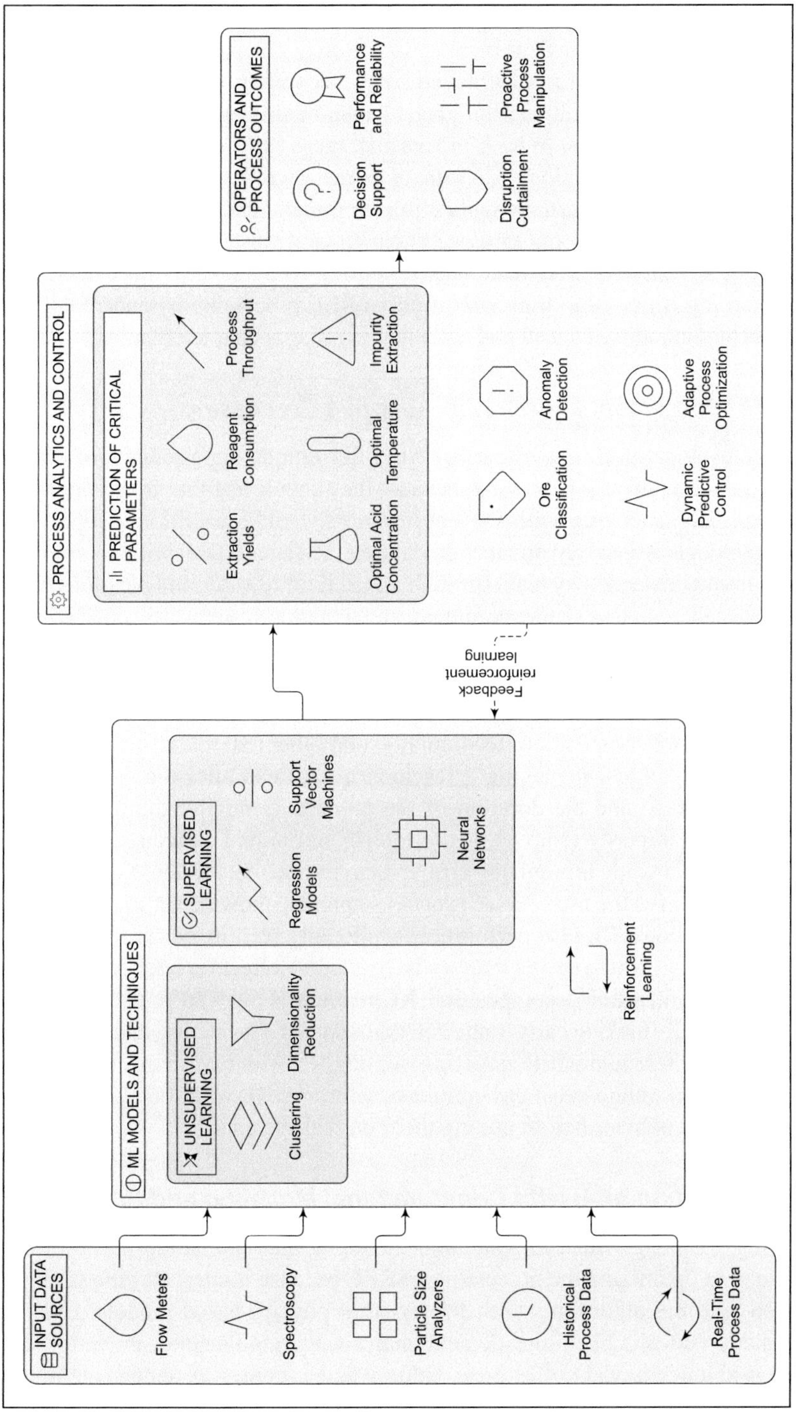

Fig. 1 Machine Learning Architecture for Predictive Analytics.

and chemical data with deep learning enhances classification accuracy and result reliability. Trace rare earth minerals detection, or separating some spectral overlapping minerals, can be achieved more accurately using a deep learning framework where ambiguities from single data modality approaches are resolved using multiple techniques. In REE processing, these skills are crucial as even the slightest change in mineralogy can impact extraction efficiency and the purity of the end product. Deep learning applications in mineral and phase recognition lift operational effectiveness and mitigate chemical and waste byproduct emissions, improving sustainability. As data and computing resources are more accessible Deep learning is poised to transform modern REE processing by enhancing data and offering unmatched speed and accuracy in mineral characterization.

5.3 Reinforcement Learning in Adaptive Processing

RL systems and algorithms as part of AIe offer unique opportunities of control and optimization in REE processing because they enable learning from experience by continually engaging with the environment. Unlike traditional supervised learning models which rely on labelled datasets, RL agents start without any form of orientation, instead, they learn through a system of rewards and penalties based on actions taken. In less stable, nonlinear, and dynamic systems, RL stands out for intricate multilevel and metallurgical processes that are hard to model completely for enumeration. With regards to REE processing, RL systems have the capability to provide automation in real-time adjustment of control parameters in relation to changes in ore composition, feed quality, and other external disturbances. For instance, an RL agent managing a leaching circuit can alter the concentration of chemicals used and the duration of the processes, monitor the efficiency of extraction and impurity removal, and establish an optimal extraction policy that maximizes yield while minimizing cost and environmental burden. Such adaptive learning minimizes the use of fixed process setpoints, allowing for more agile and resilient operations. RL also performs exceptionally well in systems with multi-stage processing, where earlier choices have a ripple effect on later performance. Through state and action decomposition, RL guarantees full workflow optimization by systematically making early-stage belt adjustments considering later stages like energy expenditure and purity targets. Coupling RL with real sensor data, digital twin simulations, and virtual environments improves RL with the ability to train agents and test policies before putting them on real systems.

5.4 Integration of AI with Computational Modeling and Simulation

Applying AI along with computational modeling and simulation is the newest technique in optimizing the processing of REE because it integrates the strengths of deep learning algorithms with the existing physics-based models. Existing computation models like finite element analysis, thermodynamic simulation, and reaction kinetics models offer great value when it comes to understanding the physical and chemical aspects of REE extraction and separation, but many of these models are limited by system complexities, uncertainties regarding input parameters, and the ever-changing nature of industrial workflows. Integrating AI, particularly ML and deep learning, into algorithms increases model accuracy, lowers computational costs, and allows for real-time decisions. AI algorithms can be trained with process data, simulations, and other output data so they can capture correlations and patterns that conventional models fail to offer. This method allows extensive simulation which can be used for scenario testing, sensitivity analysis, and optimizing processes with multifactorial enhancements while eliminating the need for elaborate physical experimentation. One of the most significant advances in modern technology is the creation of digital twins or virtual replicas of actual AI-enhanced processing plants. These replicas can observe system behavior, optimize operations as well as predict future outcomes relying on simulations. Using AI and sensor data, digital twins, unlike their predecessors, are able to outperform past standards of detection and self-correction. They are now able to swiftly change processing techniques to counter proactively detected problems. Such interactions between the real-life and the virtual world improve agility and process control accuracy. Additionally, the simulation approach with AI boosts hydrometallurgical and pyrometallurgical activities by minimizing the risks and expenses of industrial scaling from the lab through understanding spatial interplay and controlling the parameters efficiently delineated for novel routines and separation techniques.

6. AI-Enhanced Process Optimization

AI techniques likeML and deep learning are optimizing operations throughout the REE processing life cycle, combining static, human-controlled approaches into a single, more efficient, data-centric paradigm. Algorithms for each step of the REE processing cycle (sensing, lab analysis, and historical record evaluation), utilizing cutting-edge AI-techniques, achieve maximum yield with minimum energy expenditure and ecological harm. Optimization is considered in terms of achieving a balanced recovery rate, product purity, and operational cost, which entails a complex process of constant adjustment of multiple parameters, including reagent amounts, instrumentation temperature, pH values, and processing durations. Algorithms are also capable to pinpoint minute nonlinear inter-variable interdependencies which standard statistical approaches fail.

6.1 Optimization of Leaching Conditions and Yield

Leaching is a critical step in the processing of REEs, where valuable metals are selectively dissolved from ore using chemical solvents. The recovery of REEs while protecting the environment, and minimizing reagent expenditure, energy consumption, and chemical leakage requires optimizing leaching conditions. Techniques in AI, particularly ML, optimization algorithms, and AI in general, have been adept at optimizing these processes concerning interrelations cluster between many factors like acid concentration, temperature, pH, particle size, and leaching time. Determining optimal leaching parameters through trials, as is done with many conventional techniques, is extremely taxing, inefficient, and slow, especially with ever-changing compositions of mineral ores. AI models are very important in this case as they sift through numerous datasets acquired from laboratory experiments, pilot-scale tests, and real-time plant operations, identifying patterns and predicting outcomes based on the varying leaching conditions set on yield and purity for the final product. For instance, regression models and even neural networks can be used to predict the REE extraction rate which allows easier recovery adjustment for operators. Additionally, AI could drive other optimization methods like genetic algorithms or particle swarm optimization to effectively navigate different sets of leaching conditions looking to achieve maximum yield with the minimal cost and environmental impact, which is ideal. The systems outlined above can combine several objectives, such as recovery rates and the quantity of reagents used, as well as the waste produced, thus contributing to environmentally conscious processing targets. Besides, real-time sensor data combined with AI models enables leaching parameters to be tuned to changes in ore quality or other operational disturbances on the fly. This flexibility maintains constant levels of performance while minimizing resource waste. Therefore, the AI optimization of leaching features a remarkable, more economically attractive, easier-to-implement solution for REE extraction within an enhanced sustainable environmental framework of mining activities.

6.2 Solvent Extraction and Selective Separation Improvements

Solvent extraction is one of the most important operations in REE processing, which involves the separation and purification of constituent REEs from their complex mixtures. The overall purity and economic value of the final products is critically dependent on the recovery and selectivity attained through solvent extraction. Because of the inherent chemical complexity or the resemblance of the REEs to each other, optimizing the solvent systems composed of extractants, diluents, and modifiers, as well as phase ratios, contact time, and temperature is often challenging. Enhancing processes of solvent extraction using AI promises greater flexibility through more data-driven optimization approaches. AI algorithms have the capability to analyze the large volumes of data information from experiment and operational data action databases to predict the efficiency and selectivity of extraction with different solvents as well as the anticipatory building of unit operations. These models allow determining the most optimal reagent and operating conditions that will maximize selective separation of target REE and minimize the co-extraction of non-target elements. The costly test-and-trial methods are alleviated, significantly speeding up the development of processes. Sharp AI alleviates real-time monitoring and control problems related to extraction of solvents. The intelligent integration of pH, temperature, and phase compositions allows control algorithms to adjust flow rate, phase mixing, and concentrations of the solvents to ensure optimal separation. Such flexibility is applied to the changes in feedstock and other unpredicted disturbances to the process, product quality, and stabilization. Additionally, new AI techniques such as RL enable the development of adaptive systems for the separation processes that change their operational algorithms gradually over time, making the entire system more stable and resource-efficient.

6.3 Waste Minimization and Resource Recovery

The processes involving the extraction and separation of REEs result in significant waste such as tailings, chemicals, and various other by-products posing serious issues environmentally and economically. There are opportunities for improving the ecological spill from REE processes and the sustainability of the entire supply chain pertaining to these issues with regards to minimal resource recovery and waste diminishment. Implementing more sophisticated predictive AI-powered fuel elimination systems brings about possibilities for efficient tracking and accurate waste, capture measurement, and control (CEMCs). With the aid of ML, AI algorithms can analyze complex industrial cross-process datasets, and with higher-order, scrutinize analytics cross-process data to determine inefficient resource consumption and waste-generation drivers. AI increases value chain efficiency by diminishing leftover hazardous wastes in the form of spent materials with tail-end controls optimization to boost the recovery of precious commodities from waste streams. According to him, tailings and sludge are associated with REEs and other recoverable metals which, with the aid of secondary extraction techniques, can be more economically extracted at the same time, thus improving effectiveness (Vashishth et al., 2024).

6.4 Case Studies of AI-based Process Optimization

It could be noted that AI is rapidly advancing and becoming more readily available to the general public. It has most certainly been integrated with success into optimizing the processing of REEs through a multitude of case studies and their use of AI bolsters the industrial sector. The employment of ML, RL, and digital twins enhances the efficiency, and sustainability of the processes improving yield, sustainability metrics, and overall efficiency.

Case Study 1: Machine Learning Optimization in Hydrometallurgical Leaching

In a singular hydrometallurgical case study, a plant associated with processing REEs, integrated advanced ML models to further enhance the optimization of their leaching process. The leaching parameters of an acid concentration, temperature, and duration were previously determined through painstaking and expensive trial-and-error processes. The AI system was subsequently trained to predict leaching conditions which would maximize REE recovery by analyzing patterns in historical operational data. The ML model was able to uncover intricate variable relationships which would have been impossible to recover using manual systems. The predictive model enhanced the plant yield by 15% while reducing the utilization of hazardous chemicals by 10%. This optimization improved production efficiency, environmental impact, operating costs, and showcased the benefits that AI can have during critical steps in the metallurgical sequence.

Case Study 2: AI-Driven Real-Time Control in Solvent Extraction
AI was integrated into the solvent extraction circuitry of a pyrometallurgical operation to assist with the selective separation of REEs. To achieve high recovery and purity, a combination of temperature, phase contact time, and solvent flow rate need to be controlled precisely during the extraction processes. The algorithm of the AI system looped the data concerning pH, temperature, and phase compositions of the system along with their respective models for real-time data comparison. This level of control enhanced the responsiveness and adaptability to process changes and mitigated the impact of human-invoked error. The implementation of the AI control system improved the selectivity and purity of the recovered REEs, enhanced process stability, lowered reagent consumption, and reduced reliance on human supervision. Furthermore, automated process controls diminished operator interaction with the equipment, thereby enhancing product quality while decreasing operational costs.

Case Study 3: Digital Twins and Reinforcement Learning in Pilot-Scale REE Processing
This case study examined the use of digital twins and RL to optimize the separation workflow of an REE processing plant. Digital twins can simulate the actual REE processing plant, allowing the operators to undertake multiple processing simulations and forecasts without any idle processing time, thus ensuring that process alterations can be safely tested. The phase was subsequently executed to provide relevant data-driven insights, which allowed operational decisions to be gradually improved through RL, adjusting the temperature of the separator, reagent addition rates, and mixing velocities, as necessary. Through this cyclical process, optimal separation efficiencies were accomplished along with decreased byproduct waste, reduced energy consumption, and holistic improvement. This was the first time that an AI system was able to fully automate the defined metasystem for multi-objective self-optimization of precision-driven metallurgy.

7. Intelligent Monitoring and Control Systems

A considerable leap in the processing of REEs from the intelligent monitoring and control systems stems from its ability to provide real-time supervision, specific control, and dynamic management of complex metallurgical operations. Such systems utilize AI along with modern sensors and data processing technologies to track vital process variables like temperature, pH, flow, and even the flow of chemicals. These systems use sensor fusion techniques, where data from different types of sensors is put together to generate a synthesized and accurate understanding of the environment, enabling proper and timely decisions to be made.

7.1 Real-time Monitoring with AI-Integrated Sensors

Precise and sustainable processing of REEs relies heavily on real-time monitoring. AI, in combination with modern sensor technologies, has transformed the manner in which process parameters are continuously measured and assessed during metallurgical operations. AI-embedded sensors enable the collection and archiving of immense streams of real-time data like temperature, pressure, pH, chemical concentrations, and particle size distribution. In comparison to conventional sensors, smart sensors contain AI algorithms capable of preprocessing data, identifying relevant anomalies, and filtering out noise locally prior to sending relevant details to control systems. The ability to perform data processing in a local context accelerates response time and reduces the load on the central processing unit(s). In addition, real-time predictive analytics of process changes, equipment failure, and other predictive monitoring activities are performed using ML and other AI technologies, which ensures optimal conditions for functioning and proactive monitoring. This increases stability in the processes while reducing the likelihood of unpredicted downtimes due to stabilized deviations in advance. Equipped with AI, sensor networks make it possible to integrate data from various disparate sensors to create a complete picture of the processing environment. As an example, spectroscopic sensors provide information on the chemical state and the progress of reactions occurring during leaching or solvent extraction in conjunction with temperature and flow sensors. Controlling charging leaching solutions and maintaining other operating conditions is achieved by utilizing this rich multidimensional data, enabling prompt and precise control action, thus enhancing selectivity and efficiency in REE separation.

7.2 Sensor Fusion for Accurate Data Interpretation

The technique of sensor fusion is especially helpful in streams connected to REEs owing to its intricacies. This form of data aggregation becomes relevant in the context of REE where drastic physical and chemical changes occur simultaneously. In the case of REEs, it is necessary to fuse spectroscopic and chemical measurement, flow, temperature, and pressure sensors in order to achieve the required precision. Reconciling these datasets requires the use of AI algorithms that incorporate ML data fusion, also known as (meta-)ML algorithms. These algorithms work towards smoothing the noise, resolving conflicting information, and more importantly, delineating hidden structures from the multi-sourced data. Although appreciated, the set of parameters is not limited to composition of minerals within the material, concentration of the reagents, and reaction kinetics during various leaching or separation processes. Sensor fusion allows faster detection of unmonitored process changes or failures by improving the detection of process perturbations significantly.

7.3 Intelligent Control Systems in Hydrometallurgy and Pyrometallurgy

The application of Intelligent Control Systems (ICS) is transforming the extraction of REEs, both hydro-metallurgically and pyrometallurgically, by providing real-time adaptive control of intricate chemical and thermal operations. Processes subclassified under hydrometallurgy, such as leaching, solvent extraction, and precipitation, are aqueous chemistry-based and require thermal and pH metering as well as reagent and agitation scaling generally known to optimize recovery and purity, as well as cyclonic mixing recovery. In comparison, high-temperature processing (pyrometallurgy) includes roasting, smelting, and refining, which have specific control requirements such as furnace temperature, gas flow, and residence time.

Implementation of AI-powered control algorithms allows modification of set operational conditions in real-time through active sensor feedback, monitoring, and predictive modeling. In intelligent control systems for hydrometallurgy, real-time adjustment of reagent and processor timing is achieved through monitoring of chemical concentrations and reaction progression via AI sensors and sensor fusion, leading to improved chemical recovery with minimal wastage. Such processes or systems contribute to reducing operational expenditures associated with reagent overuse and less effluent containing reagents, thus protecting the environment. In the field of pyrometallurgy, intelligent control systems utilize sophisticated thermal sensors and simulations for fuel, air, and coolant management to sustain furnace temperature. Figure 2 depicts an ICS integrating sensors, data acquisition, and AI algorithms to optimize processes in hydrometallurgy and pyrometallurgy. It enables real-time monitoring, adaptive decision-making, and efficiency improvements in metal extraction operations. The use of RL and model predictive control allows the furnace to draw from thermal historical and real-time data, optimizing energy consumption while maintaining REE separation consistency. Therefore, the use of ICS in both hydrometallurgical and pyrometallurgical processes marks progress toward fully automated, accurate, and refined sustainable REE processing. These systems improve functionality as well as safety by decreasing the amount of people needed to control the system in dangerous settings, allowing for human-free monitoring during threats to timely interventions to process perturbations.

8. AI-Driven Selective Separation Techniques

Because of the chemical similarities that calcified the relationships among different sorts of REEs, selective separation of the elements is considered one of the steps in the entire processing chain that is of utmost importance yet, at the same time, very complicated. Classic separation methods such as solvent extraction and ion exchange have always been shown to rely on multiple steps while simultaneously overspending on chemicals and fostering operational, economical, and environmental destruction. To address these problems, the application of AI in selective separation

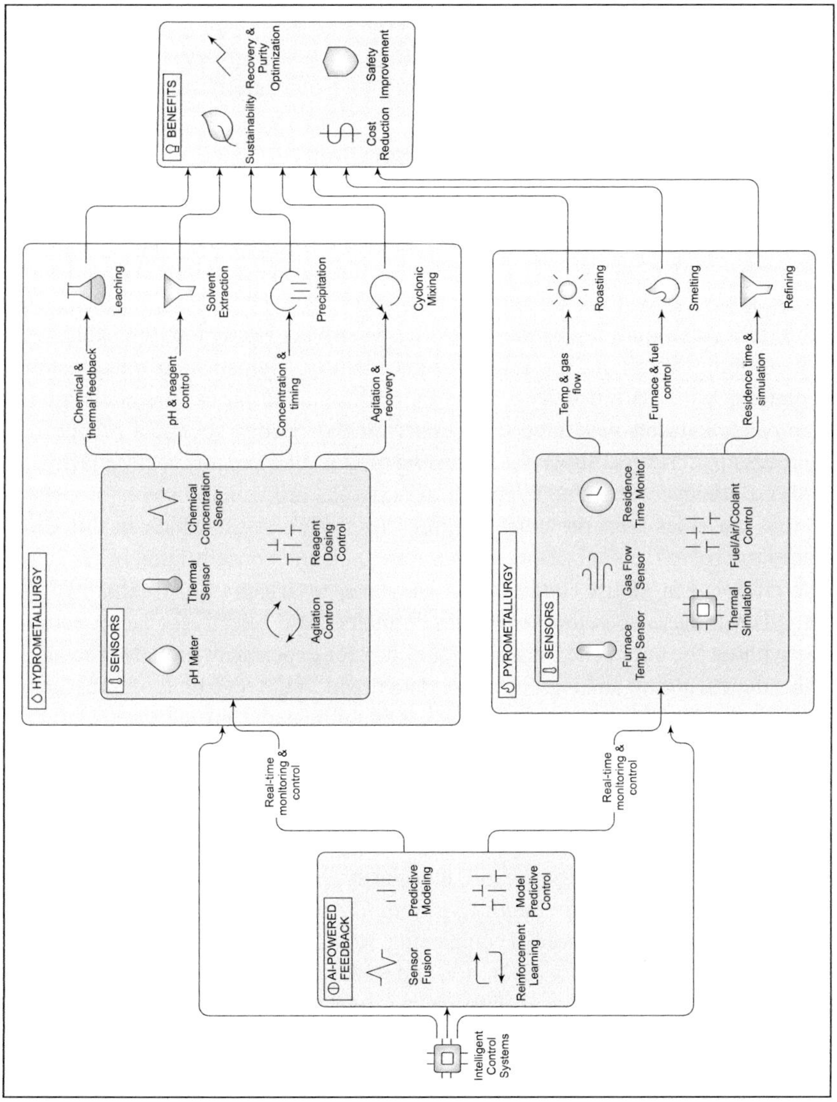

Fig. 2 Intelligent Control System for Hydrometallurgy and Pyrometallurgy.

techniques greatly improves accuracy, efficiency, and sustainable practices. These approaches use AI to change parameters such as solvents used, temperature, pH, contact time, and others to enhance REE separation. Such systems have the capability of adjusting conditions in real-time to achieve optimal separation while conserving reagents, using historical data and real-time information. In addition, AI can create new agents or extractants by scanning vast chemical databases and simulating molecular interactions, expediting the design process for advanced materials by making them more selective and environmentally friendly.

8.1 AI Algorithms for Selectivity Prediction

Precise forecasting of selectivity in the processes of separating REEs is necessary for successful and accurate extraction of individual elements from complex mixtures. Algorithms powered by AI, especially those that incorporate ML and deep learning, have proved to be effective in the modeling and prediction of selectivity for different techniques of separation including solvent extraction, ion exchange, and membrane filtration. These algorithms identify sophisticated relationships and patterns which are usually missed by traditional empirical methods due to the immense datasets generated from experimental results, chemical properties, and process parameters. Supervised learning models like support vector machines (SVMs), random forests, and even neural networks are widely used to associate the input variables with the corresponding separation result which in this case includes the solvent composition, temperature, pH, and concentration of the ions. These models can adjust conditioning relevance to selectivity of certain REEs which allows these operators to refine parameters and selectively target certain REEs without the need for extensive trial-and-error experimentation. Moreover, it enables the clustering and classifying of data to aid in the revealing of underlying patterns in the separation behavior as well as enabling some unsupervised learning techniques to aid in revealing. With advanced AI techniques such as deep RL, continuous dynamic optimization is possible by predicting changes based on real-time feedback from the separation process and updating those predictions iteratively. These algorithms adapt to changes in feed composition or process disturbances while maintaining selectivity and product purity. Moreover, AI integration with molecular simulations and cheminformatics databases allows extractants and solvents to be virtually screened, enhancing selectivity-defining characteristics. AI deployment for selectivity prediction and REE processing enhances efficiency, cost, environmental impact, and achieves the industrial benchmark of automated, environmentally friendly sustainable separation technologies.

8.2 Tailoring Process Parameters for High Purity Separation

Availability of high purity separation of REEs demands the maintenance and fine-tuning of myriad parameters like temperature, pH, solvent concentration, phase contact time, and mixing intensity. A separation process, like other processes of industrial chemistry, features a set of obstacles emanating from the conflict of factors each of which tries to optimize cost and productivity creating a heuristically solvable problem suited for automated decision-making. Most redundant processes can be solved with a new approach to AI that enables tailoring of predictive processes in real-time. AI models based on historical and real-time data streams from various separation processes are capable of predicting product purity and yield for different values of parameters. For example, ML algorithms can determine what level of adaptability for pH or solvent ratio changes is acceptable for selective chelation of certain REEs, thus uncovering more specific ranges for maximizing purity with minimal chemicals. This optimization has the potential to not only lower costs, but also reduce harmful waste, thereby promoting sustainable processing practices. Therefore, AI control systems can alter the process parameters while still in operation, giving the challenged composition and environmental conditions being fed to the system value in use. RL algorithms, for example, facilitate the construction of adaptive separation processes that learn and self-enhance over time, consistently adjusting parameter settings to ensure optimal purity outputs irrespective of the varying inputs. By automating the process parameter tailoring with AI, REE processing plants are able to achieve superior and more consistent separation outcomes. This feature is important to fulfill the high-quality standards of advanced technological uses where the purity of processed rare earth materials greatly impacts the operational efficiency and the dependability of electronic devices, magnets, and clean energy technologies.

9. Sustainability and Circular Economy Perspectives

The application of AI in sustainable and circular REE processing contributes to global initiatives aimed at the development of cleaner technology and the reduction of reliance on exhaustible natural resources, thereby strengthening the supply chain resilience. The combination of AI with sustainable practices not only serves environmental purposes, but also economically by creating new markets from materials that were previously regarded as waste and advocating constructive management of critical materials.

9.1 Reducing Energy and Chemical Usage

In order to enhance the sustainability of extraction and separation processes, it is critical to further reduce energy and chemical usage in the processing of REEs. The practices employed in traditional REE processing, including the high pyro and hydrometalurgical pyrometallurgical refinements which incorporate powerful and toxic chemical systems, are very energy intensive. These practices heavily draw water and discharge gasses like greenhouse gasses, dangerous wastes, and other pollutants, further worsening the state of the environment. However, AI offers us a more optimal solution to these issues. AI allows processes to have minimal resource consumption while maintaining yield and purity, thus achieving sustainability. ML models, for instance, are capable of identifying pathways that minimize energy consumption and chemical dosing strategies that enhance yield and purify the desired product, and then optimizing these pathways. AI algorithms also control AI-driven control systems that can seamlessly tune the temperature, pressure, and reaction time of a system in real-time to mitigate energy wastage. In pyrometallurgical and hydrometallurgical treatments, these capabilities enable AI to mitigate energy expenditure during fundamental processes. Additionally, AI facilitates the development and simulation of new low energy processing techniques and green solvents or reagents using computational chemistry and molecular modeling. AI enables the exploration of alternative formulations and process parameters that achieve the same or better separation performance with less impact on the environment by systemically studying chemical formulations and processes.

9.2 *Promoting Recycling and Secondary Resource Utilization*

The combination of the growing interest in the economic aspects of recycling and the primary mining's environmental ramifications has accentuated the focus on REEs. Extracting REEs from spent electronic devices, magnets, batteries, and industrial waste can help to lessen the dependency on primary ore extraction and aid in environmental conservation. The use of AI technologies is changing the effectiveness, accuracy, and scope of recovery processes. AI enables the precise integration of vision, sensor data fusion, and pattern recognition which allows for the automatic and exact separation of REEs from complicated waste streams. Secondary sources, in comparison to primary ores, are difficult to process because of their more complex feedstocks. Based on the chemical composition data of secondary sources, ML models dynamically adjust processing techniques and processing parameters to optimize extraction. The optimization reduces the recovery operations' chemical and energy costs, processing time, and resource expenditure (Simaei & Rahimifard, 2024). Also, variability in feedstock and contamination are some of the other challenges that processes governed by AI tend to struggle with. With the application of AI, the recycling rate and the quality of materials reclaimed through these processes are maintained persistently. By integrating technologies of AI and principles of circular economy, the management and scalability of recycling processes are improved greatly, turning former waste materials into marketable products while simultaneously closing the material loop.

9.3 *AI's Role in Life Cycle Assessment (LCA)*

As with any product—from its initial stages as raw materials to processing them into a product, and even to the use and end-of-life recycling phases, REE possesses unique challenges from an LCA standpoint that require attention. To evaluate REE's environmental impacts, attempting to assemble the data needed for evaluating energy consumption, emissions, streams of wastes, resources, and processes productivity is deeply complex and often out of scope. The application of AI is striking for the scope of environmental impact measurement needed for the life cycle assessment since it facilitates reliable, effective mechanisms for data collection, and efficient analysis of various relevant components to impact measurement. ML is capable of analyzing large quantities of data to determine vital impact elements, as well as model intricate relations within the REE processing life cycle. Consequently, different processing scenarios can be dynamically simulated, offering immediate evidence of environmental impact predictions resulting from specific operational parameter changes. AI enables emission, energy consumption, and waste generation mitigation through selective forewarning of said emissions prior to mitigation actions being taken. In addition, waste emission provides limitations AI-powered digital twins of REE processing plants facilitate the capturing and guiding of models for continuous environmental monitoring throughout the set life cycle while guaranteeing compliance with sustainability standards, eliminating the restraints of rapid mitigation adjustments to ecological footprints, and enabling swift cutbacks. The aforementioned features reveal how changes in processing conditions or materials influence overall impact of AI and assist the enhancement of green technologies. The cross-section of AI with LCA contributes to at least three aspects: fostering innovation towards greater circular and eco-efficient practices for sustaining rare earth extraction and processing; aiding in regulatory compliance and corporate sustainability reporting; and, in general, AI adapts and implements LCAs within companies in a way that changes their nature from retrospective accounting to forward-looking planning. In the case of deep learning, we can indisputably say that AI will be a powerful enabler of subsequent iterations of LCA (Vishwakarma, 2025).

10. Challenges and Future Directions

Having noted the advances AI brings to the processing of REEs considering its application in AI, there are several debilitating constraints to its impact and value which restrain me from detailing further:

1. **Data Availability and Quality:** In attempts to apply AI technology within processing systems of REEs, data scarcity compounded by lack of uniformity presents as a major challenge. The extraction of REEs faces the challenge of complex geological and chemical variation. Collecting large, accurate datasets from numerous mining sites is a challenging process. Adding to the challenge are issues of noisy, sparse, or heterogeneous data due to diverse measurement techniques, environmental conditions, and equipment standards. This combination is intrinsically undermining, due to the abundance of data posing difficulty in training AI models. The reliance on low-quality data compounded with problems of data standardization makes AI algorithms fail to optimally predict or generalize the REE processes.
2. **Integration with Existing Infrastructure:** The application of AI on existing systems faces multiple technical and operational challenges. Many of the

mining and metallurgical plants implement outdated machinery and control systems which AI technologies cannot integrate into. The infrastructure's retrofit or upgrade to enable real-time AI analytics and automated control requires a considerable financial investment and specialized knowledge. Furthermore, the AI models must maintain dependability in harsh and fluctuating industrial settings, managing near-instantaneous data streams in real-time. The smooth integration of AI components with standard processed units of older technology systems poses intricate challenges, often needing bespoke systems and sophisticated engineering.

3. **Explainability and Trustworthiness:** One major concern on the REE processing AI is model interpretability for metallurgical processes with potential high danger risks of financial drains, safety risks, and environmental impact. Many advanced AI models such as deep learning networks function as "black boxes" and issue products without lucid reasoning. This AI model interpretability issue poses a need for explanation for loss of trust for engineers, operators, and regulators who would need to validate and verify decisions made through AI prior to wide-scaling through-frame reconsidered acceptance. Developing explainable AI (XAI) techniques that provide insights into model reasoning is essential, but remains challenging due to the complex, nonlinear nature of metallurgical processes and AI algorithms.
4. **Cybersecurity Risks:** The expanding connectivity of AI powered systems in REE processing through digital networks and IoT devices increases their susceptibility to cyber threats. Cyber threats may include attempts to compromise AI algorithms, manipulate control systems, or alter data repositories leading to unsupervised, uncontrollable, and hazardous processing conditions, or proprietary information theft and espionage. Implementing strong cybersecurity defenses such as secure communications, attack prevention, and passive surveillance responsiveness is hasty, demanding prepositive responsiveness and constant adaptation to new, emerging challenges. The technology has to be actively structured to respond adaptively to changing threats. Risks, not mitigated, can inappropriate operational dependability and trust in the AI system's dependability.
5. **Regulatory and Ethical Concerns:** Applying AI technologies to process REE raises boundaryless pertinent moral debates. There exist no normative instructions within the limits of AI application in mining and metallurgy law fostering gaps of shirked accountabilities. Issues related to the commodification of data, ownership, and control, governance of AI, and fair and equitable distribution of value are all unresolved. Other ethical issues include the nearly unrestrained potential consequences of AI enhanced eco-terrorism. This kind of illegal activity targets vulnerable areas, particularly those close to mining operations. Responding to these questions requires the deployment of AI into societal frameworks guided by clearly defined policy frameworks.

The merging of AI techniques into traditional metallurgical science REE processing is expected to yield breakthroughs towards intelligent, sustainable adaptive operations—this will define the new paradigm we hope to achieve. Following are some of the future directions:

1. **Hybrid AI and Physics-Based Modeling:** AI has promising models for the future, particularly in hybrid models which combine data-driven approaches with traditional physics-based simulations as these can have far-reaching benefits. They capitalize on the strength of both methods, where physical laws guide AI predictions, while AI modifies its algorithms to account for complex exercises found in the real world. This amalgamation allows one to circumvent challenges created due to limited or uncertain data, and enhances the overall precision, dependability, and interpretability of a process model. Such models are more helpful when optimizing and controlling the separation and extraction of the REE as an aid in more sophisticated operations.
2. **Advanced Sensing and Real-Time Data Integration:** Combining AI with advanced sensors will transform REE processing by allowing real-time monitoring and control. The invention of IoT coupled smart sensors, high-precision tenable and non-dispersive spectroscopy, and other advanced sensor fusion techniques will produce abundant, uninterrupted streaming data of chemical, physical, and environmental parameters. AI algorithms can analyze the data in real-time for anomaly detection, dynamic optimization of process parameters, automated predictive maintenance, and much more. All of this culminates in faster reaction times and less downtime, along with unparalleled insight into using the available resources sustainably.
3. **Digital Twins and Virtual Process Optimization:** In the upcoming applications of AI, digital twin technology—which acts as virtual duplicates for the actual REE processing plants—will be tremendously significant. AI-driven digital twins equipped with real-time data have the potential to emulate numerous simulations in processing, try out changes without actual operation halting, and anticipate results under set parameters. With this capability, constant virtual optimization, experimenting without risks, and acceleration of innovation while attending to the environment, operational costs, and risks become possible. The development of digital twins will be crucial for flexible, responsive, and smarter REE processing facilities.
4. **AI-Driven Circular Economy and Recycling Innovations:** Further research is bound to zero in on AI's influence on folding circular economy principles into the rare earth industry. Utilizing AI, recycling methods can be optimized to reclaim REEs from obsolete products, mining garbage, and industrial discards, thus reducing primary extraction dependency and minimizing damage to Mother Nature. ML models will devise strategies for the efficient parting of materials' sections and processing conditions for secondary resources, which will strengthen sustainable supply lines. These innovations will be essential to address the scarcities of resources while ensuring nature's eco-friendly equilibrium (Quinto et al., 2025).

5. **Workforce Development and Collaborative Human-AI Systems:** More advanced technology will allow for higher reliance on human-AI interfaces within REE processing where AI serves as a supplement as opposed to the primary intelligence. Workforce development involving management of automated systems will require an AI decision-support tool that is user-friendly to operators and engineers. There will be a shift in training courses designed for active professionals towards enabling them to navigate and operate advanced automated systems, enhancing socioeconomic inclusivity. Along with these advancements will be the framework to ensure that the integration of new technology fosters growth (Vashishth et al., 2024).

11. Conclusion

Embedding AI into REE processing provides a more efficient, accurate, and eco-friendly extraction and separation approach. AI techniques including ML, deep learning, and RL facilitate predictive modeling, real-time analytic monitoring, automated control, and resource optimizing in separation processes which decreases energy usage, waste generation, and increases recovery of resources. These changes boost the operational efficiency and cost-effectiveness of the organization while improving environmental social governance (ESG) and circular economy objectives. AI has a solution for the restrictions posed by traditional approaches to techniques and will bring the industry to a more intelligent, sustainable, adaptable future, especially during the growing times of REEs in clean energy, electronics, and defense industries.

References

Adisyahputra, Irkham, Pratomo, U., Rochani, S. and Hartati, Y.W. (2024). Advancement in Ion-Imprinted Polymers for Separation and Analysis of Rare Earth Elements. *Separation and Purification Reviews*, 1–20.

Ahmed, I., Seku, K., Nidamanuri, S.B. and Koyyala, K.K. (2025). Handling metals waste to salvage with automation. In: *Harnessing Automation and Machine Learning for Resource Recovery and Value Creation*, pp. 187–212. Elsevier.

Bashynska, I. and Prokopenko, O. (2024). Innovative Technologies and Digital Platforms: AI's Role in a Sustainable Circular Economy. *Scientific Journal of Bielsko-Biala School of Finance and Law, 28*(3).

Chittalury, D. (2025). Chemical Breakthroughs Facilitating Superior Material Reclamation Techniques. In: *AI Technologies for Enhancing Recycling Processes*, pp. 101–112. IGI Global Scientific Publishing.

Delogu, M., Berzi, L., Dattilo, C.A. and Del Pero, F. (2023). Definition and sustainability assessment of recycling processes for bonded rare earths permanent magnets used on wind generators. *Advances in Materials and Processing Technologies*, 9(2), 608–654.

Kesar, B. and Joseph, S. (2025). Impact of E-Waste Processing on Health and Environment. In: *Integrated Approaches for Sustainable E-Waste Management*, pp. 283–308. IGI Global Scientific Publishing.

Liang, G., Liang, Y., Niu, D. and Shaheen, M. (2024). Balancing sustainability and innovation: The role of artificial intelligence in shaping mining practices for sustainable mining development. *Resources Policy*, 90, 104793.

Patel, D.K., Sharma V.K. and Sharma H.O. (2024). Advancements in aluminum matrix composites: A comprehensive review on the mechanical and microstructural attributes of rare earth elements (REEs) and rare earth oxides (REOs) doped materials. *Transactions of the Indian Institute of Metals*, 77(9), 2265–2276.

Patel, D.K., Sharma V., Sharma H.O. and Pardeep Kumar. (2024). Application of AI in Material Science to Accelerate Material Innovation. In: *Artificial Intelligence in Material Science*, pp. 198–207. CRC Press.

Patel, D.K., Sharma V.K. and Sharma H.O. (2025). Current global Status of rare earth elements (REEs) and their role as catalysts in reducing air pollution for maintaining environmental sustainability. *Transactions of the Indian Institute of Metals*, 78(1), 4.

Patel, D.K. and Sharma V.K. (2025) Illuminating Innovations: A Comprehensive Review of Rare Earth-Doped Phosphors and Their Applications in LEDs. *Transactions of the Indian Institute of Metals*, 78(7), 165.

Quinto, S., Law, N., Fletcher, C., Le, J., Antony Jose, S. and Menezes, P.L. (2025). Exploring the E-Waste Crisis: Strategies for Sustainable Recycling and Circular Economy Integration. *Recycling*, 10(2), 72.

Rojas, L., Peña, Á. and Garcia, J. (2025). AI-Driven Predictive Maintenance in Mining: A systematic literature review on fault detection, digital twins, and intelligent asset management. *Applied Sciences*, 15(6), 3337.

Sharma, M., Gorti J., Sharma V.K., Kumar V. and Joshi R.S. (2022). Comparative prediction of surface roughness for MAFM finished aluminium/silicon carbide/aluminium trioxide/rare earth oxides (Al/SiC/Al2O3)/REOs) composites using a Levenberg–Marquardt Algorithm and a Box–Behnken Design. In: *Proceedings of the Institution of Mechanical Engineers, Part E: Journal of Process Mechanical Engineering*, 236(3), 790–804.

Sharma, V.K., Kumar, P., Akhai, S., Kumar, V. and Joshi, R.S. (2024). Analyzing the tribological and mechanical performance of Al-6061 with rare earth oxides: An experimental analysis. In: *Proceedings of the Institution of Mechanical Engineers, Part E: Journal of Process Mechanical Engineering*, 238(5), 2278–2292.

Sharma, V.K., Kumar, V., Joshi, R.S. and Kumar, A. (2022). Effect of REOs on tribological behavior of aluminum hybrid composites using ANN. In: *Additive Manufacturing in Industry 4.0*, pp. 153–167. CRC Press.

Sharma, V.K., Kumar, V. and Joshi, R.S. (2021). Manufacturing of stable hydrophobic surface on rare-earth oxides aluminium hybrid composite. In: *Proceedings of the Institution of Mechanical Engineers, Part E: Journal of Process Mechanical Engineering*, 235(4), 899–912.

Simaei, E. and Rahimifard, S. (2024). AI-based decision support system for enhancing end-of-life value recovery from e-wastes. *International Journal of Sustainable Engineering*, 17(1), 80–96.

Singh, B., Ahmed, S., Chandra, S. and Singh, A. (2025). Scaling Green Hydrogen Production through AI Innovations: Review of Production Technologies and Global Policies Impact. *Leveraging AI for Innovative Sustainable Energy: Solar, Wind, and Green Hydrogen*, pp. 395–408. IGI Global.

Srikant, Sharma V.K., Kumar V., Joshi R.S., Jain S. and Kumar P. (2021). A review of recent research on rare earth particulate composite materials and structures with their applications. *Transactions of the Indian Institute of Metals*, 74(11), 2569–2581.

Vashishth, T.K., Sharma, V., Agarwal, S., Gupta, N., Singh, S. and Sharma, R. (2024). Embracing the Current Challenges in Sustainable Waste Management in Tourism and Hospitality. In: *Hotel and Travel Management in the AI Era*, pp. 219–256. IGI Global.

Vashishth, T.K., Sharma, V., Sharma, K.K., Kumar, B., Chaudhary, S. and Panwar, R. (2024). Waste-to-Energy Solutions Harnessing IoT and ML for Sustainable Power Generation in Smart Cities. In: *Methodologies, Frameworks, and Applications of Machine Learning*, pp. 125–146. IGI Global.

Vashishth, T.K., Sharma, V., Sharma, K.K., Kumar, B., Chaudhary, S. and Panwar, R. (2024). Environmental sustainability and carbon footprint reduction through artificial intelligence-enabled energy management in electric vehicles. In: *Artificial Intelligence-Empowered Modern Electric Vehicles in Smart Grid Systems*, pp. 477–502. Elsevier.

Vashishth, T.K., Sharma, V., Sharma, K.K., Kumar, B., Panwar, R. and Chaudhary, S. (2024). Efficiency Assessment and Optimization in Renewable Energy Systems Using Data Envelopment Analysis. In: *Data Envelopment Analysis (DEA) Methods for Maximizing Efficiency*, pp. 223–244. IGI Global.

Vashishth, T.K., Sharma, V., Sharma, K.K., Kumar, P. and Joshi, R.P. (2024). Leveraging Blockchain Technology for Sustainable Development Goals: A Comprehensive Overview. *Driving Transformative Technology Trends with Cloud Computing*, 278–308. IGI Global.

Vishwakarma, S.K. (2025). Circular Economy in Aerospace: Recycling Composites and Rare Metals. *International Journal of Management and Business Development*, 2(05), 20–37.

Yin, Y. and Yang, Y. (2025). Sustainable Transition of the Global Semiconductor Industry: Challenges, Strategies, and Future Directions. *Sustainability*, 17(7), 3160.

Zhang, L., Chen, H., Pan, J., He, X., Shi, S., Long, X. and Zhou, C. (2025). Partitioning Behavior of Lithium and Rare Earth Elements During Coal Preparation and Their Differences in Mode of Occurrence. *Mineral Processing and Extractive Metallurgy Review*, 1–12.

3 Role of Rare Earth Elements in Electric Vehicles and Sustainable Energy and Mobility Solutions

Ranjana Singh[1*], Subhash Chand[2] and Prachi Bhatnagar[3]

1. Introduction

The global transition to sustainable energy systems has positioned electric vehicles (EVs) as a cornerstone in reducing greenhouse gas emissions and combating climate change. As governments, industries, and consumers increasingly prioritize decarbonization, EVs have emerged as a transformative technology, offering an environmentally friendly alternative to internal combustion engine (ICE) vehicles (International Energy Agency [IEA], 2023). However, the production of EVs relies heavily on a group of critical materials known as rare earth elements (REEs), which are indispensable for the performance and efficiency of key components such as electric motors and batteries. These 17 chemically similar elements, including the 15 lanthanides plus scandium and yttrium, possess unique magnetic, luminescent, and catalytic properties that make them essential in high-tech and clean energy applications (Srikant et al.,2021; Sharma et al., 2024; Patel et al., 2024). Despite their name, REEs are not inherently rare in the Earth's crust, but their economically viable deposits are geographically concentrated, and their extraction and processing pose significant environmental and geopolitical challenges (Alonso et al., 2021). This article explores the pivotal role of REEs in EV manufacturing, their applications, supply chain dynamics, and the implications for the future of sustainable mobility, with a focus on fostering a circular economy and reducing dependency on primary sources.

REEs are critical to the functionality of EVs, particularly in the production of high-performance permanent magnets used in electric motors. Neodymium-iron-boron (NdFeB) magnets, which often incorporate elements like neodymium, praseodymium, dysprosium, and terbium, are favored for their exceptional magnetic strength and resistance to demagnetization, enabling compact, lightweight, and efficient motors that enhance EV performance (Stanford Materials Corporation, n.d.). For instance, these magnets allow EVs to achieve higher torque and power output, contributing to longer ranges and better acceleration. Additionally, REEs

[1] Department of Commerce and Management, IIMT University, Meerut (UP), India.
[2] Department of Mechanical Engineering, IIMT University, Meerut (UP), India.
[3] Department of Computer Science & Engineering, IIMT University, Meerut (UP), India.
* Corresponding author: singh4ranjana@gmail.com

DOI: 10.1201/9781003683254-3

such as lanthanum and cerium are used in nickel-metal hydride (NiMH) batteries, which were prevalent in early hybrid vehicles, and in catalysts that improve battery performance in certain lithium-ion configurations (Liu et al., 2018; Patel et al., 2024). The demand for REEs in EVs is projected to grow significantly, with studies estimating that automotive electrification in China alone could require up to 315 thousand tons of REEs by 2030, driven by the increasing adoption of hybrid EVs (HEVs), plug-in hybrid EVs (PHEVs), battery EVs (BEVs), and fuel cell EVs (FCEVs). This surge in demand underscores the strategic importance of REEs in the global energy transition.

However, the reliance on REEs introduces complex challenges related to supply chain security, environmental sustainability, and geopolitical risks. China dominates global REE production, accounting for over 60% of the world's supply, which creates vulnerabilities due to potential export restrictions and price volatility (Patel et al., 2025; Sehgal et al., 2025; Patel et al., 2025). For example, posts on X have highlighted concerns about China's near-monopoly on rare earth magnets, noting that 99% of these critical components are produced in China, posing risks to global automotive supply chains. The extraction and processing of REEs are also energy-intensive and environmentally damaging, releasing toxic and sometimes radioactive waste, which complicates efforts to align EV production with sustainability goals. These challenges have spurred efforts to develop recycling technologies, alternative materials, and diversified supply chains to mitigate risks. For instance, innovations like Tesla's development of rare earth-free permanent magnet motors and the exploration of tetrataenite as a substitute for NdFeB magnets demonstrate the industry's push toward reducing REE dependency (Tesla, 2023; Rollat et al., 2018) investigated alternative battery technologies, emphasizing solid-state and lithium-sulfur batteries as solutions to decrease reliance on REEs in EVs. Their research highlights that these advanced battery systems can achieve higher energy density while reducing exposure to raw material supply risks. They estimate that REE demand for automotive electrification could reach 315,000 tons by 2030, with neodymium and dysprosium accounting for 51% and 20%, respectively. Additionally, China controls approximately 70% of raw REE production and over 90% of refined REEs, creating significant supply chain and policy risks (U.S. Geological Survey, 2024); explored methods for recycling and recovering REEs from end-of-life EV components. Their research highlights that closed-loop recycling can alleviate raw material shortages, decrease reliance on primary sources, and support sustainable expansion of electric vehicle production. Zhou et al., 2020); note that the surging demand for EVs significantly increases the need for REEs, particularly neodymium (Nd) and dysprosium (Dy). The International Energy Agency (IEA, 2021) forecasts a rise in the global EV fleet from 3.1 million in 2017 to 125 million by 2030, potentially escalating neodymium demand from 582–1,162 tons in 2017 to 20,000–40,000 tons by 2030, and dysprosium demand from 180–360 tons to 6,000–13,000 tons over the same period. IEA, 2021 conducted a comparative analysis of REE-free motor technologies, finding that ferrite and synchronous reluctance machines can match the performance of traditional REE-based motors in many scenarios, though they may experience reductions in power density and torque. Li et al. (2020) explain that Nd-Fe-B magnets, which contain a lot of neodymium, help EVs have more power and travel farther. This makes the vehicles more energy-efficient, allows for smaller batteries, and helps them compete better in the clean-energy market. Ramesh et al. (2020) explored the addition of dysprosium to Nd magnets to improve their stability at high temperatures. This ensures consistent motor performance under demanding conditions, which is essential for high-performance EVs to operate safely across various temperatures. Policy strategies were examined and adopted by the European Union and the United States to decrease dependence on REEs. Their study emphasizes incentives for recycling, advancements in alternative materials, and global partnerships to diversify supply chains. Wilson et al. (2021) suggest creating magnets with reduced REE content while maintaining strong performance. By using composite structures and advanced processing techniques, this method decreases reliance on raw materials, lowers costs, and supports a more sustainable EV industry moving forward.

2. The Role of Rare Earth Elements in EV Components

2.1 Permanent Magnets for Electric Motors

The most significant application of REEs in EVs is in the production of permanent magnet synchronous motors (PMSMs), which rely on NdFeB magnets containing neodymium, praseodymium, dysprosium, and terbium. These magnets provide superior magnetic strength, allowing for smaller, lighter motors with higher efficiency compared to induction motors (Stanford Materials Corporation, n.d.). For example, a typical EV motor may contain 1–2 kg of REEs, contributing to improved torque density and energy efficiency, which are critical for extending driving range (Alonso et al., 2012). Dysprosium and terbium are particularly important for enhancing the magnets' resistance to high temperatures, ensuring performance stability in demanding conditions.

- Neodymium (Nd): Provides the base for high-strength permanent magnets.
- Dysprosium (Dy) and Terbium (Tb): Improve thermal resistance, making motors more durable.

These magnets enable the compact and lightweight designs of EV motors, translating to higher energy efficiency and extended driving ranges. Sustainability of Permanent Rare Earth Magnet motor in H (EV) industry as shown in Figure 1.

2.2 Batteries and Energy Storage

Batteries and Catalysts

While lithium-ion batteries dominate modern EVs, REEs like lanthanum and cerium play a role in certain battery chemistries and catalysts. Lanthanum is used in NiMH batteries, which are still found in some hybrid vehicles like the Toyota Prius,

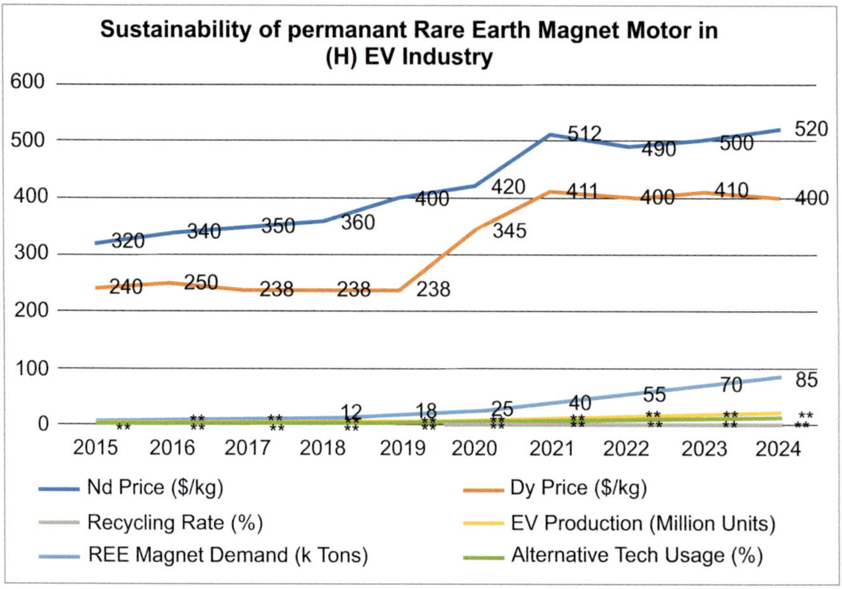

Fig. 1 Sustainability of Permanent Rare Earth Magnet Motor in H (EV) industry (Li et al., 2020).

due to their durability and cost-effectiveness. Cerium is employed as a catalyst in battery manufacturing processes to improve electrochemical performance. Although the proportion of REEs in batteries is smaller compared to motors, their contribution to battery efficiency remains significant (Stanford Materials Corporation, n.d.).

- Lanthanum (La): Used in NiMH batteries, particularly in hybrid vehicles.
- Lithium and Nickel: Though not REEs, these materials are often used alongside REEs in battery systems.

3. Power Electronics and Lighting

REEs also contribute to the efficiency and durability of EV power electronics. Cerium and lanthanum are used in catalysts and polishing processes for semiconductor manufacturing, while yttrium finds application in LED lighting, crucial for vehicle displays and external lighting systems.

- Cerium (Ce): Used in polishing compounds for electronic components.
- Yttrium (Y): Enhances energy efficiency in lighting systems.

4. Supply Chain Dynamics and Geopolitical Challenges

China's control over more than 60% of global REE production and 99% of rare earth magnet manufacturing creates significant supply chain vulnerabilities

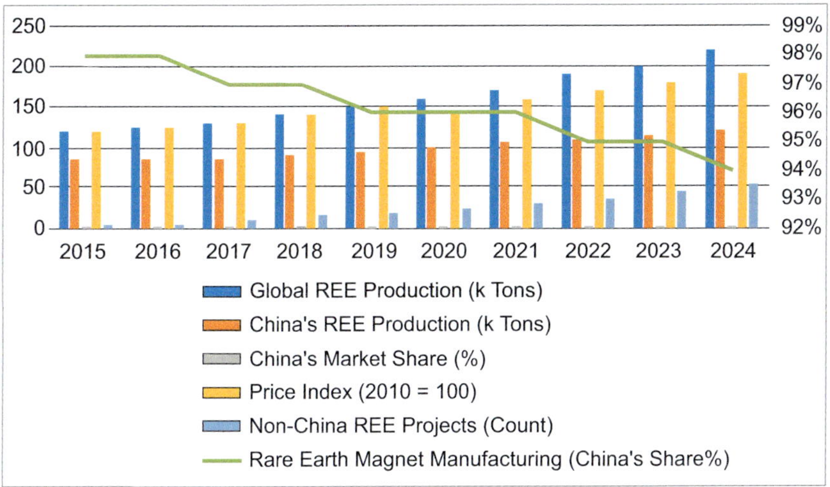

Fig. 2 Global Production and China's Dominance.

(Victory Metals Australia, 2023). Historical events, such as China's 2010 export restrictions, led to price spikes and prompted global efforts to diversify supply sources. Countries like Australia, the United States, and Canada are investing in domestic REE mining and processing, but scaling these operations to meet EV demand remains a challenge (IEA, 2023). Global Production and China's Dominance is shown in Figure 2.

5. Environmental Impact of REE Extraction

REE mining and processing are associated with severe environmental degradation, including soil erosion, water contamination, and radioactive waste generation. For example, the Bayan Obo mine in China, a major REE source, has been linked to significant ecological damage. These impacts raise concerns about the sustainability of EV production, as the environmental footprint of REE extraction may offset some of the climate benefits of EVs (Figure 3).

6. Geopolitical Risks

The concentration of REE supply in China poses geopolitical risks, particularly amid trade tensions and export controls. Recent posts on X have emphasized the strategic importance of securing alternative REE sources to reduce dependency on Chinese exports. Efforts to establish supply chains in regions like North America and Europe are underway, but these initiatives require substantial investment and time (IEA, 2023).

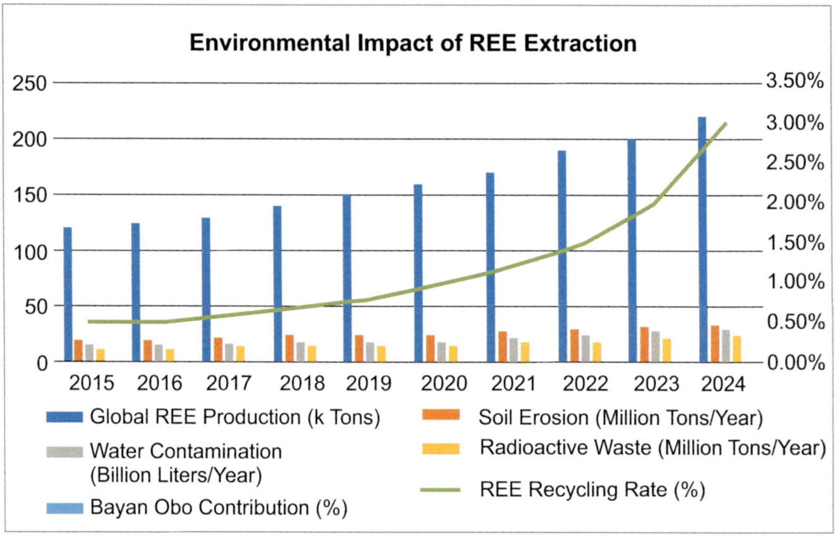

Fig. 3 The environmental impact of REE extraction (U.S. Geological Survey, 2024).

6.1 Demand and Supply Dynamics

The global rise in EV adoption has driven significant demand for REEs. The International Energy Agency (IEA, 2021) projects the EV fleet to grow from 3.1 million in 2017 to 125 million by 2030, increasing demand for neodymium and dysprosium. Estimates suggest neodymium demand could rise from 582–1,162 tons in 2017 to 20,000–40,000 tons by 2030, with dysprosium demand increasing from 180–360 tons to 6,000–13,000 tons (Table 1). In China, the largest EV market, it is predicted that REE demand for automotive electrification could reach 315 thousand tons by 2030, with neodymium and dysprosium comprising 51% and 20%, respectively.

China dominates the global REE supply, producing about 70% of raw REEs and over 90% of refined materials, raising concerns about supply security. Recent export restrictions, including those on terbium in April 2025, highlight vulnerabilities, particularly amidst U.S.-China trade tensions. Efforts to diversify supply include U.S. initiatives like MP Materials' California mine and Australian projects such as Lynas Corporation's Texas facilities. However, these face challenges in matching China's cost advantages and infrastructure, as noted by the U.S. Geological Survey, 2024. Demand and supply of REE is shown in Figure 4.

6.2 Environmental and Social Challenges

REE extraction and processing are environmentally intensive, often generating toxic and radioactive waste. Du and Graedel (2011) highlight that China's lax environmental regulations contribute to its dominance but result in significant

Table 1 Global REE supply and demand for 2017 and 2030.

Metric	*2017*	*2030 (Projected)*
EV Fleet (Million Units)	3.1	125
Neodymium Demand (tons)	582–1,162	20,000–40,000
Dysprosium Demand (tons)	180–360	6,000–13,000
Automotive REE Demand in China (thousands of tons)	—	315
Neodymium (%) of Automotive REE Demand in China	—	51
Dysprosium (%) of Automotive REE Demand in China	—	20
China's Market Share in Global REE Supply (raw)	—	70
China's Market Share in Global REE Supply (refined)	—	90

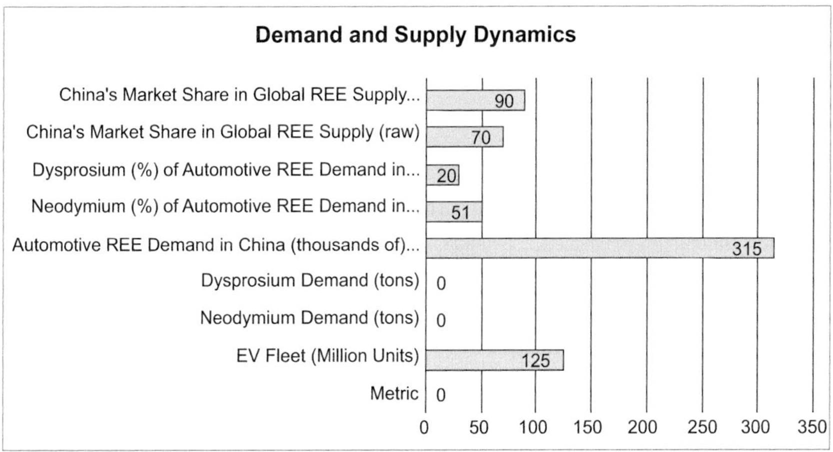

Fig. 4 Shows the demand and supply of REE (U.S. Geological Survey, 2024).

ecological damage. These impacts challenge the sustainability of EVs, as upstream environmental costs offset downstream emission reductions. Social issues, including labor exploitation in mining regions, further complicate the ethical landscape of REE sourcing.

Recycling offers a potential solution suggesting that end-of-life EV motors could become a significant source of secondary REEs, potentially surpassing U.S. domestic mining output by 2050. However, recycling NdFeB magnets is complex due to their dispersed use and energy-intensive processes, with current recycling rates remaining low.

6.3 *Innovations to Reduce REE Dependence*

Reducing REE reliance is a key focus for researchers and industry. Wilson et al. propose designing magnets with lower REE content while maintaining high performance, using composite structures and advanced processing techniques to reduce raw material dependency, lower costs, and promote a more sustainable EV industry. Tesla (2023) aims to develop REE-free permanent magnet motors, potentially using ferrite-based alternatives, having already reduced REE usage by 25% in its Model 3 drive units from 2017 to 2022. Other automakers, like BMW, are adopting externally excited synchronous machines (EESMs) that eliminate REE-based magnets.

Emerging technologies include soft magnetic composites (SMCs) for motor cores and carbon nanotube (CNT) windings for improved conductivity. Startups like Niron Magnetics, supported by General Motors and Stellantis, are developing REE-free magnets. Rollat et al. (2016) emphasize that advancements in battery technologies, such as solid-state or lithium-sulfur batteries, could further reduce REE reliance in EV production.

Policy and Economic Implications

Policy interventions are essential to address supply challenges. The U.S. has prioritized domestic REE production, but competing with China requires substantial investment. Policies promoting recycling, such as battery passports for material tracking, could enhance sustainability, as suggested by Filippas et al. (2021), Vipin et al. (2019) Vipin et. al. (2021), and Dinesh et al. (2025). Economic analyses indicate that REE price volatility, driven by China's export policies, impacts EV manufacturing costs. A 2010 export quota led to a 30% drop in global REE production, highlighting market sensitivity.

7. Conclusion

EVs are revolutionizing transportation, cutting emissions, and boosting efficiency. REEs are crucial for EV motors, batteries, and electronics. However, surging REE demand creates challenges like scarcity, supply chain risks, and price volatility, worsened by concentrated production in China. Innovations like REE-free magnets, composite materials, alternative batteries, and externally excited machines aim to reduce dependency. Collaboration among industry, policymakers, and researchers is essential for a resilient EV ecosystem. The future of EVs relies on balancing performance with sustainable resource management through innovation, supportive policies, and global cooperation to ensure long-term sustainability.

References

Alonso, E., Sherman, A.M., Wallington, T.J., Everson, M.P., Field, F.R., Roth, R. and Kirchain, R.E. (2012). Evaluating rare earth element availability: A case with revolutionary demand from clean technologies. *Environmental Science and Technology,* 46(6), 3406–3414. https://doi.org/10.1021/es203518d.

Du, X. and Graedel, T.E. (2011). Global in-use stocks of the rare earth elements: A first estimate. *Journal of Industrial Ecology,* 15(6), 836–843. https://doi.org/10.1111/j.1530-9290.2011.00362.x.

Filippas, A., Smith, B. and Johnson, C. (2021). Future usage of permanent magnets in electric vehicles: A lifecycle perspective. *Sustainability,* 13(16), 8923. https://doi.org/10.3390/su13168923.

International Energy Agency. (2021). *The Role of Critical Minerals in Clean Energy Transitions*. https://www.iea.org/reports/the-role-of-critical-minerals-in-clean-energy-transitions.

Kumar, D., Sharma, V.K. and Sharma, H.O. & quot. (2025). Current global Status of rare earth elements (REEs) and their role as catalysts in reducing air pollution for maintaining environmental sustainability, & quot. *Transactions of the Indian Institute of Metals*, 78(1), 1–11.

Kumar, V., Kumar, V. and Joshi, R.S. & quot. (2019). Investigation of rare earth particulate on tribological and mechanical properties of Al-6061 alloy composites for aerospace application & quot. *Journal of Materials Research and Technology*, 8(4), 3504–3516.

Kumar, V., Kumar, V. and Joshi, R.S. & quot. (2021). Manufacturing of stable hydrophobic surface on rare-earth oxides aluminium hybrid composite & quot. In: *Proceedings of the Institution of Mechanical Engineers, Part E: Journal of Process Mechanical Engineering*, 235(4), 899–912.

Li, J., Zhang, H. and Liu, W. (2020). Performance enhancement of Nd-Fe-B magnets in electric vehicle motors. *Journal of Magnetism and Magnetic Materials,* 512, 167012. https://doi.org/10.1016/j.jmmm.2020.167012.

Liu, Y., Zhang, J. and Chen, X. (2018). Pathways for automotive electrification and rare earth element demand in China. *Energy Policy,* 117, 206–215. https://doi.org/10.1016/j.enpol.2018.03.005.

Patel, D.K., Kumar, S., Sharma, V.K., Sharma, H.O., and Prasad, A. (2024). Alternative Source Systems of In-Vehicle Electricity Production. *Sustainable Mobility: Policies, Challenges, and Advancements*, 61–78.

Patel, D.K., Sharma, V.K., Kumar, V., Kumar, P. and Kumar, D. (2025). Enhancing hydrophobicity and anti-corrosion properties of Al-6061Aluminum hybrid composites through trace additions of rare earth oxides. In: *Proceedings of the Institution of Mechanical Engineers, Part E: Journal of Process Mechanical Engineering*, 09544089251317381.

Patel, D.K., Sharma, V.K. and Sharma, H.O. (2024). Advancements in aluminum matrix composites: A comprehensive review on the mechanical and microstructural attributes of rare earth elements (REEs) and rare earth oxides (REOs) doped materials. *Transactions of the Indian Institute of Metals*, 77(9), 2265–2276.

Patel, D.K. and Sharma, V.K. (2025). Illuminating Innovations: A Comprehensive Review of Rare Earth-Doped Phosphors and Their Applications in LEDs. *Transactions of the Indian Institute of Metals*, 78(7), 165.

Ramesh, P., Kumar, S. and Rao, V. (2020). Dysprosium-enhanced Nd magnets for high-temperature electric vehicle applications. *Materials Science and Engineering: B, 261*, 114672. https://doi.org/10.1016/j.mseb.2020.114672.

Rollat, M., Singer, D.A. and Deshaies, M. (2016). Prospective analysis of rare earth element demand in Europe. *Resources Policy,* 49, 292–301. https://doi.org/10.1016/j.resourpol.2016.06.011.

Sehgal, A., Sharma V.K., Gautam P.S. and Partap A. (2024). Forging the Future: Exploring Aluminium AA-6061 Composites through Stir Casting for Advanced Applications and Future Innovations. *Evergreen – Joint Journal of Novel Carbon Resource Sciences and Materials*, 806–826.

Sharma, V.K., Kumar, P., Akhai, S., Kumar, V. and Joshi, R.S. (2024). Analyzing the tribological and mechanical performance of Al-6061 with rare earth oxides: An experimental analysis. In: *Proceedings of the Institution of Mechanical Engineers, Part E: Journal of Process Mechanical Engineering*, 238(5), 2278–2292.

Srikant, Sharma V.K., Kumar, V., Joshi, R.S., Jain, S. and Kumar, P. (2021). A review of recent research on rare earth particulate composite materials and structures with their applications. *Transactions of the Indian Institute of Metals*, 74(11), 2569–2581.

Tesla. (2023, March 1). *Investor Day 2023: Rare Earth-free Motor Development*. Electrek. https://electrek.co/2023/03/01/tesla-rare-earth-free-motor/.

U.S. Geological Survey. (2024). *Mineral Commodity Summaries 2024: Rare Earths*. https://www.usgs.gov/centers/nmic/rare-earths-statistics-and-information.

Wilson, A., Brown, D. and Taylor, R. (2021). Low-REE magnet designs for sustainable electric vehicle motors. *Journal of Sustainable Metallurgy, 7*(3), 987–998. https://doi.org/10.1007/s40831-021-00392-5.

Zhou, L., Wang, X. and Li, Y. (2020). Electric vehicles and greenhouse gas emission reductions: The role of rare earth magnets. *Environmental Research Letters, 15*(10), 104023. https://doi.org/10.1088/1748-9326/aba5f3.

4 Rare Earth Elements in Sustainable Catalysis

Bhaskar Garg[1*], Tanuja Bisht[2] and Yong-Chien Ling[3]

1. Introduction

The demand for sustainable chemical technologies has grown substantially over the last two decades in response to the global need to reduce greenhouse gas emissions, minimize waste generation, and conserve finite resources. Central to this transformation is the field of *sustainable catalysis*, which aims to develop catalytic processes that are not only efficient and selective, but also environmentally benign and economically viable throughout their life cycle (Anastas & Warner, 2000; Sheldon, 2018; Poliakoff et al., 2002; Patel et al., 2025). Catalysts are the cornerstone of modern chemical industries, enabling over 90% of all chemical manufacturing processes globally (Ertl, 2008). Within this context, *rare earth elements* (REEs) have emerged as vital components in the development of next-generation catalysts that meet the stringent demands of green chemistry and process intensification.

REEs, comprising the 15 lanthanides along with scandium (Sc) and yttrium (Y), possess a unique combination of physicochemical properties, including unfilled 4f orbitals, high coordination numbers, flexible oxidation states (notably Ce^{3+}/Ce^{4+} redox pairs), and high thermal stability, that distinguish them from transition metals, traditionally used in catalysis (Bünzli, 2015; Bünzli & Pecharsky, 2013; Patel et al., 2025). These characteristics make REEs particularly attractive for applications ranging from acid-base catalysis to redox and photoredox catalysis, across a variety of reaction environments.

The catalytic versatility of REEs has been exploited in several energy- and environment-relevant transformations, including water-gas shift reaction, and CO_2 methanation due to their strong basic sites and thermal robustness (Rosid et al., 2019; Özdemir, Çiftçioğlu, & Öksüzömer, 2023). In this book chapter, we highlight the importance of REEs in various industrial and environmental sectors,

[1] Department of Chemistry, School of Basic Sciences and Technology, IIMT University, Meerut 250001, Uttar Pradesh, India.
[2] Department of Chemistry, Indira Priyadarshini Government Girls P. G. College of Commerce, Haldwani 263139, Uttarakhand, India.
[3] Department of Chemistry, National Tsing Hua University, 101, Section 2, Kuang-Fu Road, Hsinchu 30013, Taiwan.
* Corresponding author: bhaskargarg111@gmail.com, bhaskargarg_sobst@iimtindia.net

DOI: 10.1201/9781003683254-4

which are currently influencing catalysis research at the interface of chemical science, materials science, and nanotechnology, and offering a compelling avenue to simultaneously advance chemical performance while meeting the sustainability demands of the 21st century (Dagwar, Iqbal, & Dutta 2025; Tunsu et al., 2025; Cherkezova-Zheleva, 2024).

2. Properties of Rare Earth Elements

2.1 Elementary Idea about Rare Earth Elements

REEs form a group of 17 chemically similar elements, comprising the 15 lanthanides (atomic numbers 57–71), along with Sc and Y (Asadollahzadeh, Torkaman, & Mostaedi, 2020) (Figure 1). The lanthanides are characterized by the progressive filling of the 4f orbitals. However, due to insufficient shielding of these inner f-electrons, the effective nuclear charge increases across the series, leading to a systematic reduction in ionic radii, commonly referred to as the "lanthanide contraction" (Owens et al., 2019). Based on their electronic configuration (Table 1) and distinct physicochemical behaviour, the lanthanides (excluding Sc) are often subdivided into two groups: light REEs) such as La, Ce, Pr, Nd, Pm, Sm, and Eu; and heavy REEs, including Gd, Tb, Dy, Ho, Er, Tm, Yb, Lu, and Y. This classification not only influences their magnetic and catalytic behaviour but also impacts their natural abundance and separation strategies.

Table 1 Valence Shell Electronic Configuration of REEs.

REEs	*Atomic Number*	*Valence Shell Electronic Configuration*
Sc	21	[Ar] $3d^1 4s^2$
Y	39	[Ar] $4d^1 5s^2$
La	57	[Xe] $4f^0 5d^1 6s^2$
Ce	58	[Xe] $4f^1 5d^1 6s^2$
Pr	59	[Xe] $4f^3 5d^0 6s^2$
Nd	60	[Xe] $4f^4 5d^0 6s^2$
Pm	61	[Xe] $4f^5 5d^0 6s^2$
Sm	62	[Xe] $4f^6 5d^0 6s^2$
Eu	63	[Xe] $4f^7 5d^0 6s^2$
Gd	64	[Xe] $4f^7 5d^1 6s^2$
Tb	65	[Xe] $4f^9 5d^0 6s^2$
Dy	66	[Xe] $4f^{10} 5d^0 6s^2$
Ho	67	[Xe] $4f^{11} 5d^0 6s^2$
Er	68	[Xe] $4f^{12} 5d^0 6s^2$
Tm	69	[Xe] $4f^{13} 5d^0 6s^2$
Yb	70	[Xe] $4f^{14} 5d^0 6s^2$
Lu	71	[Xe] $4f^{14} 5d^1 6s^2$

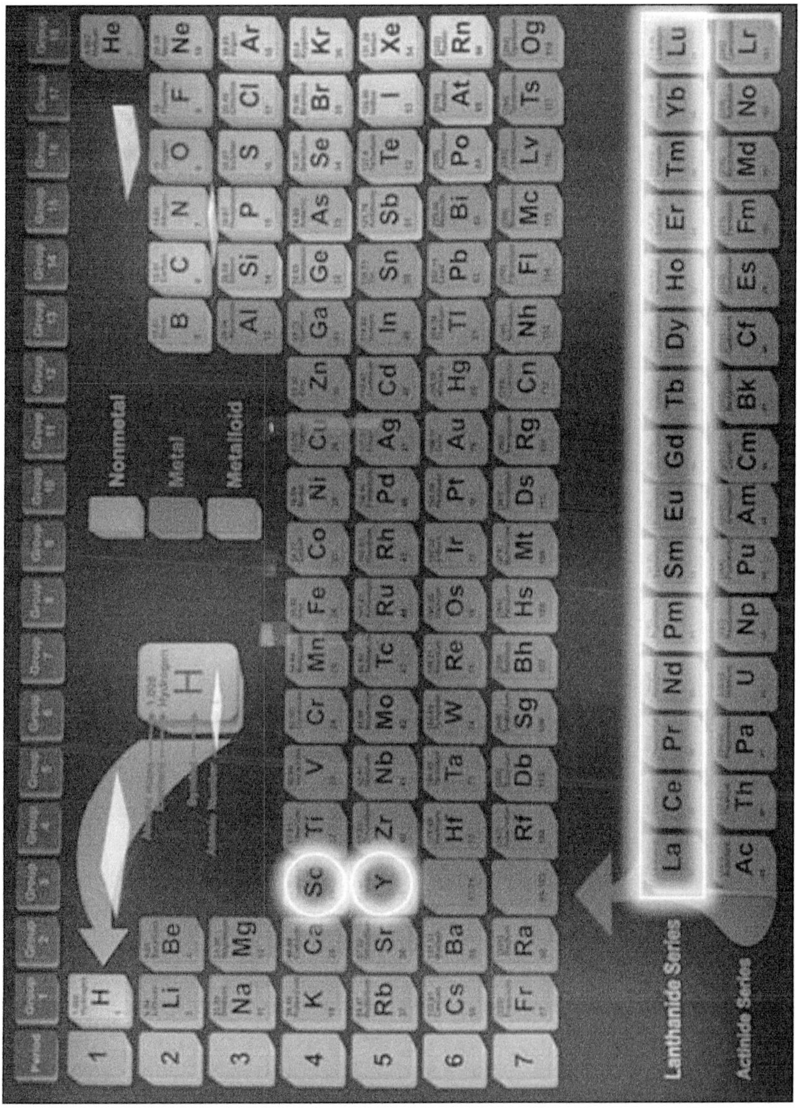

Fig. 1 Visualization of Rare Earth Elements, Glowing White, within the Periodic Table.

In terms of geological distribution, REEs occur in the Earth's crust at concentrations of up to 153 g per tonne, with Ce being the most abundant. Their natural distribution tends to follow the Oddo-Harkins rule, wherein elements with even atomic numbers are more prevalent than their odd-numbered neighbours. REEs are commonly found in three mineralogical forms: (i) as integral components in the crystal lattice of primary RE minerals such as monazite and bastnaesite; (ii) as substitutional ions replacing Ca, Sr, Ba, Mn, or Zr in minerals like apatite or europium titanite; and (iii) as surface-adsorbed ions on clays or within intergranular spaces of secondary minerals such as mica. Typically, they occur as trivalent cations forming ionic compounds, which in solid-state matrices are coordinated within polyhedral environments. These fundamental characteristics play a pivotal role in defining the reactivity, separability, and catalytic adaptability of REEs in advanced applications.

2.2 *Oxidation States and Coordination Chemistry of REEs*

REEs, particularly the lanthanides, exhibit rich coordination chemistry that arises from their large ionic radii, high electropositive nature, and predominantly trivalent oxidation state. The most stable and commonly encountered oxidation state for nearly all lanthanides is +3. This is due to the relative ease with which they lose their outer $6s^2$ and $5d^1$ electrons, forming Ln^{3+} cations with a $[Xe]4f^n$ configuration. However, certain lanthanides can access additional oxidation states, for instance, +2 and +4, depending on their electronic configurations and the nature of the ligands and reaction medium.

Ce is the most well-known lanthanide capable of stabilizing a +4-oxidation state under ambient conditions, typically in the form of CeO_2, which plays a central role in redox catalysis and oxygen storage applications. The Ce^{3+}/Ce^{4+} redox couple is particularly important in catalytic converters and other oxidative processes, where facile electron transfer contributes to catalytic efficiency. Other lanthanides such as Pr, Tb, and Dy can also exhibit +4 oxidation states. However, these REEs are relatively less stable and generally require oxidizing environments. In contrast, Eu, Sm, and Yb can adopt a +2 oxidation state, particularly in reducing or solid-state environments. The +2 state is stabilized due to half-filled (Eu^{2+}: $4f^7$) or fully filled (Yb^{2+}: $4f^{14}$) 4f subshells, which offer additional thermodynamic stability. The redox flexibility of these elements has led to their integration into catalytic systems involving single-electron transfer, photoredox catalysis, and multivalent ion batteries (Liang et al., 2022).

As stated above, the coordination chemistry of REEs is primarily governed by their large ionic size and electrostatic interactions rather than crystal field stabilization. As a result, REE cations are classified as hard Lewis acids and favour binding with hard bases such as oxygen- and nitrogen-containing ligands. Coordination numbers typically range from 6 to 12, with the most common being 8 or 9. Unlike transition metals, where d-orbital participation dictates geometry, REE complexes often adopt flexible, irregular geometries due to the more spherical and shielded nature of the 4f orbitals. Furthermore, lanthanide contraction significantly influences coordination behaviour. Heavier lanthanides, with smaller radii, tend to form more compact coordination spheres and exhibit stronger Lewis acidity. This affects ligand binding affinity, exchange rates, and overall complex stability, which can be strategically harnessed in ligand design and catalytic applications (Shannon, 1976).

Overall, the versatile redox states and adaptable coordination environments of REEs open up a wide array of possibilities in both homogeneous and heterogeneous catalysis, enabling them to participate in redox cycles, stabilize high-energy intermediates, and activate small molecules under mild conditions.

3. Role of REEs in Catalytic Transformations

3.1 REEs as Catalysts in Wastewater Treatment

Wastewater consists of a variety of organic dyes, pharmaceutical chemicals, pesticides and other industrial chemicals as the major pollutants. These pollutants significantly affect human health and the natural eco-system (Ortúzar et al., 2022; Sundararaman et al., 2022). The conventional methods of wastewater treatment often suffer with certain disadvantages. For example, pharmaceutical residues and synthetic dyes are structurally stable and often pass through conventional wastewater treatment plants (WWTPs) unaltered, leading to their continuous discharge into water bodies. This demands for highly efficient alternative methods for the effective treatment of wastewater. In this context, advanced oxidation processes (AOPs) have emerged as a powerful tool to mitigate the problems associated with conventional methods of wastewater treatment. In particular, AOPs rely upon the generation of a variety of reactive oxygen species (ROS), in particular, ·OH.

Owing to their Lewis acidic properties, an appealing redox activity, high surface area, ease of processing in both water and aqueous-organic solvents, recoverability, and reusability, REEs have attained significant reputation in the regime of green chemistry. In particular, REEs as catalysts can promote the formation of ROS upon exposure to UV light in a variety of photocatalytic reactions, including ozonation and Fenton-like systems, influencing the advanced oxidation processes significantly for the treatment of wastewater.

3.1.1 REEs as Catalysts in Degradation of Dyes

Dyes are among the most prevalent and persistent organic pollutants found in industrial wastewater, particularly from textile, leather, paper, and dye-manufacturing industries. These synthetic colourants resist fading under light, heat, and chemical exposure, making them difficult to degrade through conventional biological or physical treatment methods. Even at low concentrations, dyes can impart intense colouration to water bodies, reducing light penetration and affecting aquatic photosynthesis. Many dyes and their breakdown products are toxic, carcinogenic, or mutagenic to aquatic organisms and humans. Azo dyes, in particular, are known to degrade into aromatic amines under anaerobic conditions, which pose serious health hazards. The presence of dyes also increases the chemical oxygen demand (COD) and total organic carbon of wastewater. Their recalcitrant nature necessitates AOPs to ensure effective removal.

While the catalytic role of REEs, especially as composites, in dye degradation has been significantly explored across diverse studies, it is beyond the scope of this chapter to present an exhaustive review. Instead, selected examples of notable and impactful contributions have been included to illustrate the most relevant advancements in this rapidly evolving area (Table 2).

A La-Fe montmorillonite (La-Fe MMT) composite was prepared via a straightforward precipitation method and evaluated for the degradation of methylene blue (MB) and rhodamine B (RhB) under both batch and fixed-bed modes (Fida et al., 2017). In particular, the system achieved over 97% and 96% removal of MB and RhB (each at 100 mg/L) within 1 h in batch studies. In a fixed-bed reactor, it maintained 91% MB degradation and 65% COD removal even after 200 h. Fluorescence and quenching studies identified ·OH as the dominant reactive species. It was demonstrated that the Incorporation of La improved dyes degradation at neutral pH and reduced Fe leaching, thereby extending catalyst longevity. Owing to its cost-effectiveness, strong activity, reusability, and low leaching, La-Fe MMT could be a promising heterogeneous F-L catalyst for dye-contaminated wastewater treatment.

A series of Cu-loaded zeolite Y catalysts was developed to investigate their effectiveness in the heterogeneous F-L degradation of the Congo red (CR) dye (Singh, Rekha, & Chand, 2016). Among the catalysts, the one containing 7.5 wt% Cu demonstrated the highest performance, achieving 93.58% degradation, 95.34% decolorization, and 79.52% mineralization after 2.5, 2, and 4 hours, respectively. Additionally, this catalyst maintained consistent activity over three successive cycles.

Recently, $PrFeO_3/CeO_2$ nanocomposites (NCs) with varying CeO_2 content (0–18 wt%) have been synthesised via solution combustion followed by thermal treatment in air (Seroglazova et al., 2023). PXRD and EDXS analyses confirmed the formation of two distinct phases: orthorhombic $PrFeO_3$ (13.6–36.6 nm) and cubic CeO_2 (7.4–11.5 nm). Morphological and surface analyses such as SEM, TEM, and BET revealed foam-like, mesoporous structures of as-synthesized NCs with surface areas ranging from 8.8 to 19.4 m^2/g. Aside from that UV–Visible diffuse reflectance showed visible-light absorption (band gap 2.07–2.13 eV). The NCs exhibited enhanced adsorption (Freundlich isotherm, $Kf = 45.7$ at 15 wt% CeO_2) and superior photo-F-L activity (rate constant 0.056 min^{-1} at 9 wt% CeO_2) for methyl violet (MV) degradation under visible light. A synergistic effect of CeO_2 addition was observed, and the materials demonstrated good cyclic stability and magnetic separability, indicating strong potential for wastewater remediation.

Praseodymium-doped cadmium tungstate (Pr-$CdWO_4$) nanoparticles (NPs) were synthesised and utilized as sonocatalysts for degrading the azo dye, Remazol Black B (RBB) (Ahmadi et al., 2020). Under optimal conditions (pH 3, 100 mg/L dye, 0.35 g/L catalyst, 298 K, 100 min, 60 kHz ultrasound), 93.9% degradation was achieved. The presence of radical scavengers (Na_2SO_4, Na_2CO_3, NaCl) reduced degradation efficiency, while enhancers (KIO_4, $K_2S_2O_8$, H_2O_2) slightly improved it, with H_2O_2 yielding 98.7% degradation. Compared to individual sonolysis (34.7%) and adsorption (39.5%), the combined sonocatalytic approach was significantly more effective. Kinetic analysis showed that the pseudo-first-order model best described the process ($R^2 = 0.993$). These findings highlight the potential of Pr-$CdWO_4$ nanocatalysts in efficient sonocatalytic treatment of dye-contaminated wastewater.

In yet other work (El-Kholy et al., 2023), the erbium ion (Er^{3+})-doped TiO_2 (Er^{3+}/TiO_2) NCs have been developed to explore the visible-light-driven photocatalytic degradation of MB. Specifically, pure TiO_2 and Er^{3+}/TiO_2 NCs were synthesized via the sol-gel method as depicted in Figure 2. The as-synthesized NCs were suitably characterized using different spectroscopic tools such as FTIR, HR-SEM, EDX, XRD, XPS, BET, zeta potential, and particle size analysis. Various operational parameters were optimized, including pH, flow rate, catalyst dosage, NP ratio, dye concentration, and aeration. Whereas pure TiO_2 showed 85% MB degradation under UV light, Er^{3+}/TiO_2 NCs achieved about 77% degradation under visible light at pH 5. Interestingly, the photocatalytic efficiency increased to 80% at a flow rate of 3 L/h. Nevertheless, higher dye concentration (30 mg/L) reduced degradation to 70%. These results indicate the effectiveness of Er^{3+}/TiO_2 NCs as a visible-light-active photocatalyst for dye removal.

Graphene, a groundbreaking carbon-based nanomaterial discovered in 2004, and graphene-based materials have captivated global scientific attention due to its exceptional physicochemical properties, including high surface area, and tunable surface functionalities. These attributes make graphene and its derivatives an ideal platform for a wide range of catalytic applications, particularly in heterogeneous catalysis, where it serves as an excellent support material for metal NPs, enhancing catalytic activity, stability, and electron transfer efficiency (Garg & Ling, 2013; Garg, Bisht, & Thomas, 2017; Garg, Bisht, & Ling, 2016). Additionally, graphene-based materials exhibit intrinsic catalytic activity in several oxidation and acid-catalysed reactions, making them valuable in sustainable organic syntheses, biosensing, and environmental remediation (Garg, Bisht, & Ling, 2014 a, 2014b, 2015; Garg & Bisht, 2016; Garg & Ling, 2018; Garg, 2018).

Given the versatile electrical properties and tunability of graphene-based systems, a ternary composite was developed by integrating visible-to-UV upconversion property of β-NaYF4:Ho^{3+} with reduced graphene oxide (rGO) and TiO_2 (Fan, Wu, & Wu, 2017). The as-developed composite, β-$NaYF_4$:Ho^{3+}@ TiO_2-rGO, could absorb visible light at 450 nm, emitting UV light at 290 nm. The lanthanide-doped ternary NC could be used as an effective photocatalyst in the degradation of RhB dye with an improved efficiency (92%) within 10 h

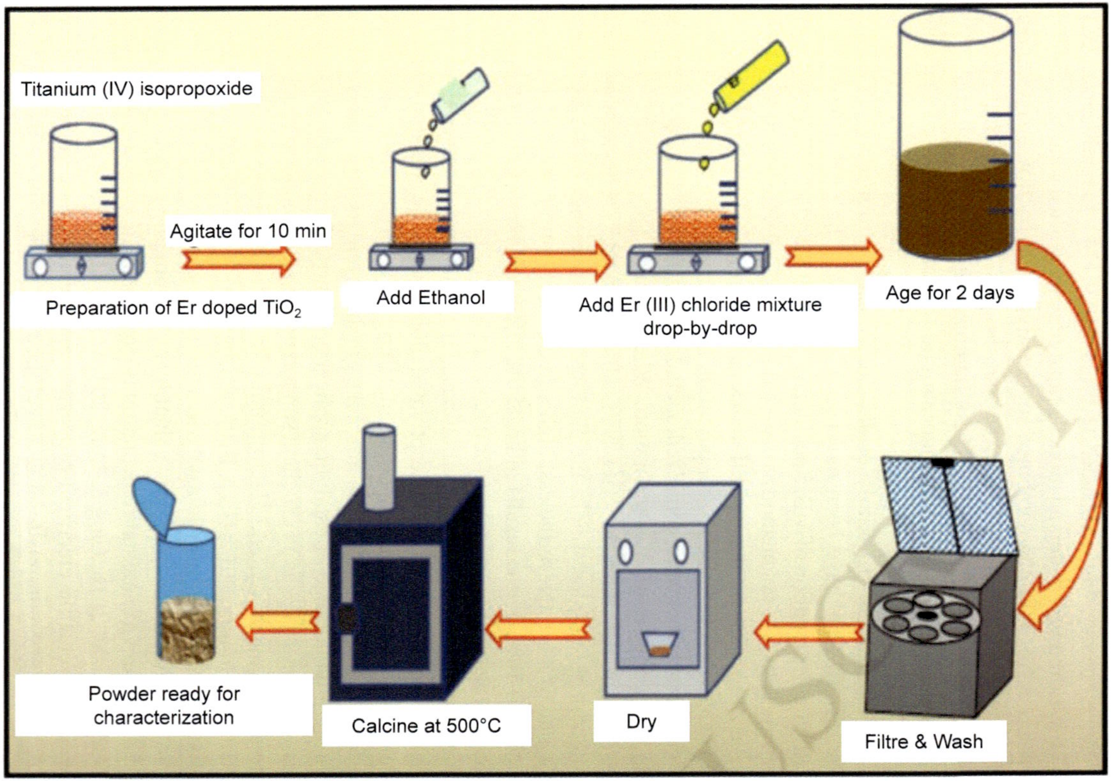

Fig. 2 A schematic, showcasing the preparation of (Er^{3+})-doped TiO_2 NCs via sol-gel method. Reproduced from El-Kholy et al. (2023).

as shown in Figure 3. The enhanced photocatalytic activity of β-$NaYF_4$:Ho^{3+}@TiO_2-rGO was attributed to relatively the high dye adsorption and electron-hole separation efficiency due to the synergistic interactions between β-NaYF4:Ho^{3+}, rGO and TiO_2. The presence of rGO not only increased charge mobility but also suppressed the recombination of charge carriers, thereby prolonging the lifetime of photo-generated electrons. This lanthanide-doped graphene-based ternary NC could be effectively recycled and reused up to four consecutive runs without significant loss in the photocatalytic activity, indicating excellent stability for practical applications.

Recently, a rGO-Dy_2O_3 NC photocatalyst was developed using a solvothermal method, where dysprosium oxide (Dy_2O_3) nanorods were uniformly distributed over rGO nanosheets (Suresh, Radhakrishan, & Sivasamy, 2024). The composite was characterized extensively and exhibited a high specific surface area of 114 m^2/g and a band gap of 3.18 eV. Under solar light irradiation, rGO-Dy_2O_3 demonstrated excellent photocatalytic performance by degrading 98.1% of MB (10 ppm) at neutral pH within 60 minutes. The reusability tests confirmed the stability of the catalyst, retaining over 90% efficiency after three consecutive cycles. The role of ·OH in the photodegradation mechanism was validated through EPR and scavenger studies. The green synthesis method, high activity, and recyclability make this nanocomposite a promising photocatalyst for real wastewater treatment.

In yet other work, Dysprosium-doped cobalt-nickel ferrite (D-CNFO) and its graphene-integrated composite (D-CNFO@G) were developed as efficient photocatalysts for the mineralization of azo dyes (Kousar et al., 2024). The synthesis was carried out using a cost-effective combination of wet-chemical and ultrasonication techniques. Structural and morphological analyses confirmed the formation of a cubic spinel phase with rod-like morphology. UV-Visible spectroscopy revealed that the D-CNFO@G exhibited improved light-harvesting capability in the visible range compared to D-CNFO alone. The photocatalytic activity was evaluated using CR, where D-CNFO@G showed a 2.4% higher degradation rate than D-CNFO. Free radical scavenging experiments confirmed the central role of ·OH in the dye degradation mechanism. Notably, the magnetic nature of D-CNFO@G allowed for its easy separation and reuse, retaining over 96% efficiency even after five successive cycles. The superior performance of D-CNFO@G is attributed to the synergistic effects of rare earth doping, graphene reinforcement, and the nanoscale architecture of the photocatalyst.

3.1.2 REEs as Catalysts in Degradation of Pharmaceutical Compounds

The chemicals released by the pharmaceutical industries are quite toxic and pose serious threats to our environment and water bodies. REEs integrated with a variety of material supports provide a handle to mitigate these toxic chemicals. For example, $LaFeO_3$ was mixed with biochar using sol-gel method successfully used as a photocatalyst in the degradation of ofloxacin (Chen et al., 2022). In particular, the optimal performance of this composite was observed at pH values above 6, and

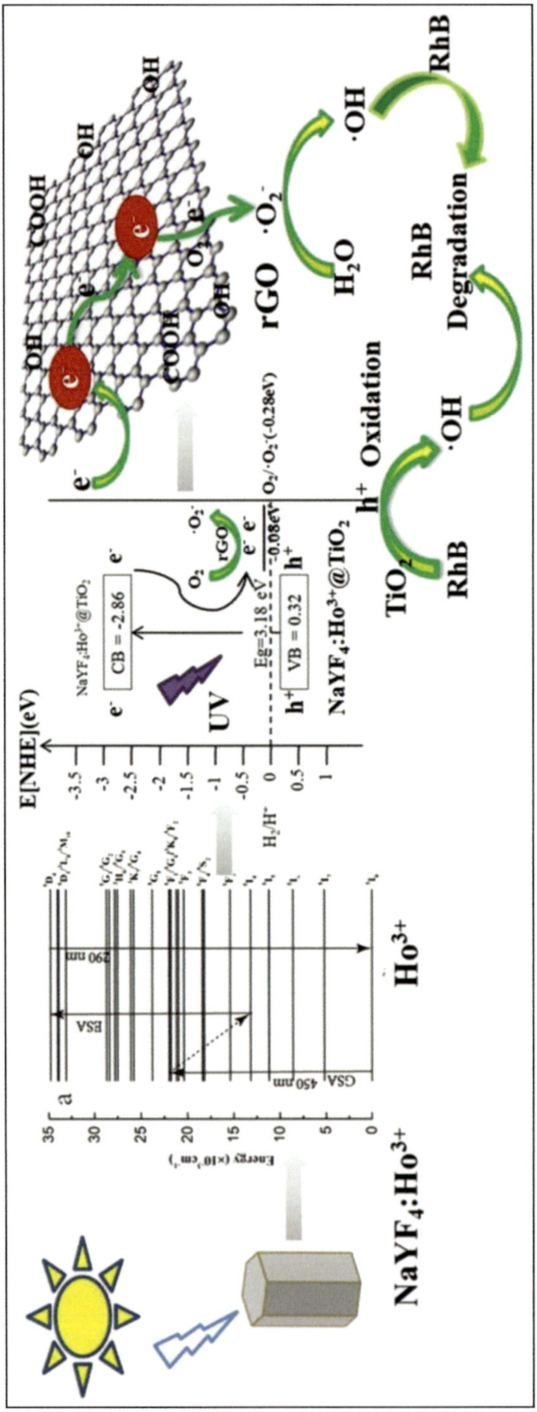

Fig. 3 Photocatalytic reaction mechanism of β-NaYF4:Ho^{3+}@TiO$_2$-rGO. Reproduced from Fan, Wu, and Wu (2017).

Table 2 REE-based catalysts for dye degradation.

Type of Catalyst	*Target Dye*	*Time (min)*	*Removal Efficiency (%)*	*References*
La-Fe MMT	RhB[a] MB[b]	60	96 97	Fida et al., 2017
Cu/Zeolite Y	CR[c]	150	93.6	Singh, Rekha, and Chand, 2016
Pr-$CdWO_4$ NPs[d]	RBB[e]	100	93.9	Ahmadi et al., 2020
Er^{3+}/TiO_2 Nanocomposites	MB[b]	60	77	El-Kholy et al., 2023
β-$NaYF_4$:Ho^{3+}@TiO_2-rGO[f] composite	RhB[a]	600	92	Fan, Wu, and Wu, 2017
rGO-Dy_2O_3	MB[b]	60	98.1	Suresh, Radhakrishan, and Sivasamy, 2024

[a]Rhodamine B. [b]Methylene Blue. [c]Congo Red. [d]Nanoparticles. [e]Remazol Black B. [f]Reduced graphene oxide.

detailed characterization linked the enhanced activity to favourable morphology, energy band structure, and photoelectrochemical properties.

Metal organic frameworks (MOFs) were also employed to integrate $LaFeO_3$, forming a $LaFeO_3$/MIL-125-NH_2 composite for enhanced photocatalytic degradation of carbamazepine and caffeine (Younes et al., 2023). Under solar simulation, the $LaFeO_3$/MIL-125-NH_2 composite exhibited superior degradation efficiency due to effective heterojunction formation and improved charge separation. Caffeine showed a higher removal (87%) than carbamazepine (74%) within 1 h. Monte Carlo simulations revealed key adsorption interactions, including metal-π, π-π stacking, and H-bonding.

Given the importance of heterojunction photocatalyst in sustainable catalysis, $Er_2FeSbO_7/BiTiSbO_6$ was developed in conjunction with several other photocatalysts, and successfully used in the degradation of enrofloxacin antibiotic over other catalytic systems (Luan et al., 2022). As such, the examples described above compellingly illustrate the remarkable efficiency and adaptability of F-L processes in targeting pharmaceutical pollutants.

3.1.3 REEs as Catalysts in Degradation of Other Wastewater Pollutants

A magnetic Fe_3O_4/CeO_2 NC was synthesised via the impregnation method and used as a F-L catalyst for degrading 4-chlorophenol (4-CP). Under optimal conditions (30°C, pH 3.0, 30 mM H_2O_2, 2.0 g/L catalyst), the system achieved a pseudo-first-order rate constant of 0.11 min^{-1} with 79.2% H_2O_2 utilization efficiency. The catalyst demonstrated good reusability over six cycles. Mechanistic studies revealed that both surface-bound and free ·OH, facilitated by Fe and Ce redox cycling and partial iron leaching, played key roles in 4-CP degradation (Xu &

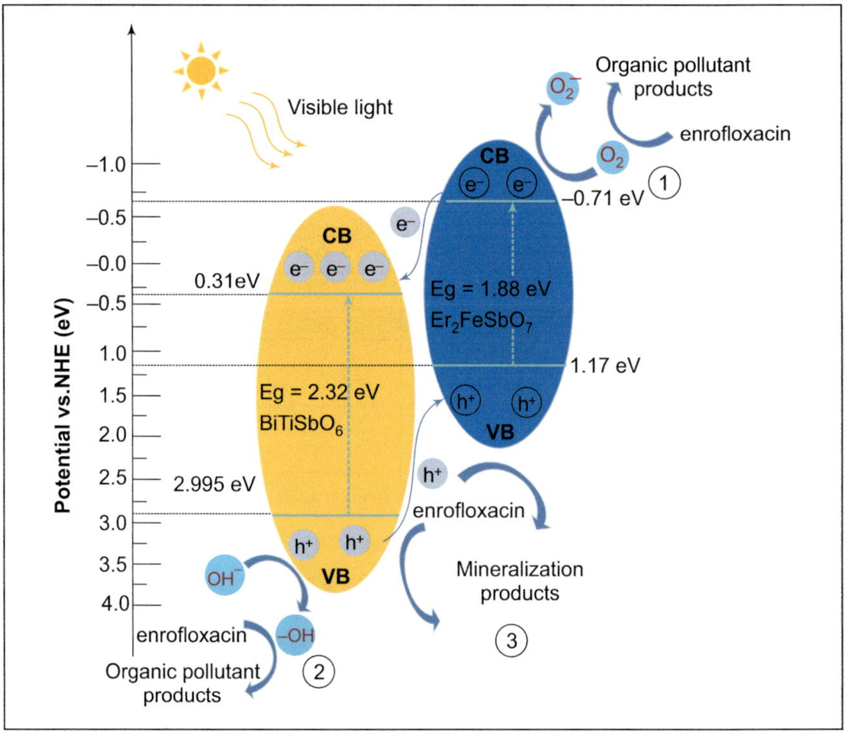

Fig. 4 Proposed photocatalytic degradation mechanism of enrofloxacin over Er_2FeSbO_7/ $BiTiSbO_7$ photocatalyst under visible light irradiation. Reproduced from Luan et al. (2022).

Wang, 2012). Separately, Gd- and Pt-doped Bi_2MoO_6 composites demonstrated enhanced photodegradation of 4-CP, attributed to the generation of ·OH, valence band holes (h^+), and superoxide radicals (Cao et al., 2015). In yet other work, Fe_3O_4/CeO_2 NC was employed in the degradation of catechol, a toxic phenolic compound.

A 2% Ce-doped ZnO photocatalyst, prepared by sonochemical impregnation showed enhanced visible-light absorption, reduced charge recombination, and efficient cyanide oxidation under UV-A and sunlight via Langmuir-Hinshelwood kinetics. It also exhibited strong antibacterial activity against *E. coli* even without light, outperforming undoped ZnO (Karunakaran, C., Gomathisankar, P., & Manikandan, G., 2010). Sm-doped ZnO nanorods (Sm/ZNRs), synthesized via a surfactant-free solvothermal method, showed enhanced visible-light absorption and red-shifted band edges compared to pure ZNRs. Under visible light, Sm/ZNRs exhibited superior photocatalytic degradation of phenol due to efficient •OH radical generation. Their 1D structure also enabled easy recyclability, indicating strong potential for environmental remediation (Sin et al., 2013). Cryptomelanes (OMS-2), synthesized via solvent-free and reflux methods, were doped with europium (Eu) through co-precipitation and ion-exchange to assess its impact on structural and catalytic properties (Cymes, Almquist, & Krekeler, 2020). Characterization confirmed that Eu doping via co-precipitation enhanced surface and catalytic features, while ion-exchange showed adverse effects. The catalytic efficiency in ethanol oxidation was highest when Eu^{3+} replaced Mn^{3+} in the solid lattice, significantly lowering activation energy (32.0 vs. 78.2 kJ/mol). The study highlights how Eu positioning governs performance. These findings offer valuable insights into optimizing cryptomelane-based catalysts through controlled rare-earth doping. Separately, a La-doped TiO_2 composite exhibited remarkable visible-light photocatalytic activity, achieving 98% degradation of NO and 38% of acetone (Ho et al., 2019). Besides, heterogeneous catalytic wet peroxide oxidation has been demonstrated as an effective approach in the photocatalytic degradation of wastewater pollutants (Fouad et al., 2021).

3.2 REEs as Catalysts in Ammonia Synthesis

Ammonia (NH_3), a key nitrogen source for fertilizers and other nitrogen-based compounds essential to agriculture, is produced via the well-known reaction, Haber-Bosch process. Although the formation of ammonia is thermodynamically favourable, the inert nature of nitrogen gas (N_2) poses a significant kinetic barrier, especially at lower temperatures. To overcome this challenge, researchers are continually exploring and developing advanced catalysts such as those promoted by REEs, especially Ru-based, that can enhance the efficiency and feasibility of NH_3 production under relatively milder conditions (Murata & Aika, 1992). For instance, a $Ru/La_{0.5}Ce_{0.5}O_{1.75}$ catalyst pre-reduced at 650°C demonstrated exceptional activity for NH_3 synthesis under mild conditions (Ogura et al., 2018). Specifically, at 350 °C, the catalyst achieved synthesis rates of 13.4, 31.3, and 44.4 mmol g^{-1} h^{-1} at 0.1, 1.0, and 3.0 MPa, respectively. The high-temperature pre-reduction led to the formation of finely dispersed Ru NPs (mean diameter <2.0 nm), partially covered by the reduced La-Ce oxide. These structural features provided thermally stable and highly active interfacial sites. Kinetic analysis confirmed the absence of hydrogen poisoning, while EELS and FTIR measurements indicated that N_2 molecules adsorbed either directly on Ru sites or interacted indirectly through the partially reduced oxide interface. In both cases, the interaction facilitated weakening of the N≡N bond, promoting its cleavage as the rate-determining step in the reaction. The mechanism, illustrated in Figure 5, emphasized the synergistic role of the Ru surface and the electron-rich oxide support in activating N_2. This work highlights how structural tuning via thermal pre-reduction could enhance N_2 activation and NH_3 production efficiency using earth-abundant oxide supports.

In yet other work, a La-promoted Ru/Al_2O_3 (Murata & Aika, 1992) demonstrated highest NH_2 production rate when the molar ratio of lanthanide to Ru was optimized at 3. Notably, the activation energies of these La-modified catalysts ranged between 46–63 kJ/mol, substantially lower than the 100 kJ/mol observed for Cs^+-promoted analogues. This superior performance was attributed to the highly dispersed lanthanide oxides at the interface of Ru particles and the Al_2O_3 support. These interactions led to a partial negative charge on the Ru species, significantly enhancing N_2 adsorption and activation, a critical step in NH_3 synthesis.

The effect of Sm on Ru/activated carbon catalyst (Sm-Ru/AC) was studied and revealed that Sm incorporation significantly enhanced Ru dispersion compared to other promoters such as Ba and K (Zhu et al., 2003). Among various formulations, the Sm-Ru/AC catalyst exhibited the highest NH_3 synthesis rate (ca. > 10-folds) attributable to the formation of partially reduced Sm_2Ox species that modulated the dispersion of active Ru sites, establishing Sm as an effective structural promoter. Furthermore, Sm was found to inhibit hydrogen adsorption and suppress carbon methanation, shifting its onset temperature from 475°C to 534°C. This resistance to methanation was attributed to the interaction between Sm and unsaturated carbon bonds, contributing to enhanced catalyst durability and performance (Zhou, Zhu, & Liu, 2010).

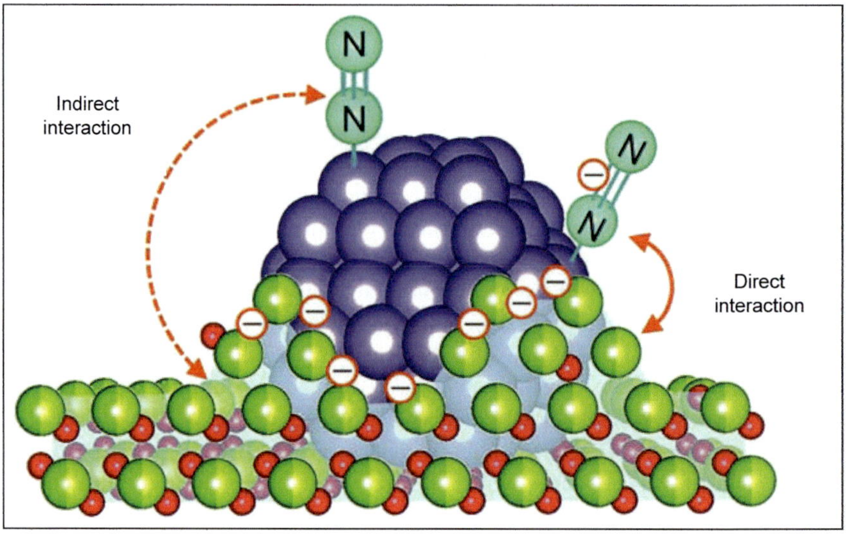

Fig. 5 The possible mechanism for of N_2 activation over $Ru/Ce_{0.5}La_{0.5}O_{1.75}$_650red. Reproduced from Ogura et al. (2018).

REs are not only effective as promoters but can also serve as efficient supports for Ru-based catalysts in NH_3 synthesis. For instance, using La_2O_3 as a support led to higher NH_3 synthesis activity compared to a system where La functioned merely as a promoter, even under atmospheric pressure (Niwa & Aika, 1996a). Interestingly, different RE oxides influenced the catalyst behaviour distinctly. While increasing the reduction temperature enhanced Ru dispersion on La_2O_3 and Sm_2O_3 supports, it had the opposite effect on CeO_{2-x}, where dispersion declined with higher reduction temperatures. This behaviour is closely tied to the reduction state of CeO_{2-x}, which, when more reduced, donates greater electron density to Ru, thereby facilitating N_2 activation. However, excessive reduction can also result in Ru aggregation or site blockage, reducing catalytic efficiency. Hence, optimizing the reduction temperature is critical to balancing the electronic benefits of reduced CeO_{2-x} with the preservation of accessible Ru sites (Niwa & Aika, 1996b). Building on these findings, a wide range of RE oxides and nitrides have since been explored in NH_3 synthesis, either individually or in combination with other oxides, further highlighting the growing significance of REEs in enhancing catalytic performance for this reaction (Li et al., 2022).

3.3 REEs as Catalysts in Petroleum Refining

In petroleum refining, fluid catalytic cracking (FCC) plays a crucial role in breaking down heavy hydrocarbons into lighter, more valuable products. In this context, Zeolites have appeared as most efficient catalysts, playing a significant role in petroleum refining.

The transition from high RE content RE-Y catalysts to low-RE-content ultrastable Y (USY) zeolites was a key development in producing lead-free gasoline. However, this shift led to a significant rise in olefin content in FCC gasoline, typically ranging from 40–45% and reaching as high as 65%, posing challenges for meeting global standards for clean fuel production. In this context, ZSM-5 zeolites were found advantageous over Y-type zeolites. For instance, when RE ions were introduced into the ZSM-5 framework via ion-exchange or precipitation methods, both olefin conversion and propylene selectivity improved significantly. The resulting gaseous products primarily consisted of propylene and C_4 olefins, with propylene yields reaching ~31%, and the olefin content in the treated naphtha decreasing to approximately 15% (Shao et al., 2009). This work underscores the potential of RE-modified ZSM-5 systems in producing cleaner fuels and more valuable light olefins. Beyond their indispensable function in zeolite-based FCC catalysts, REEs also act as vital co-catalysts in numerous other catalytic processes, dehydration, acylation, and isomerization, contributing to improved activity, selectivity, and stability (Akah, 2024).

3.4 REEs in the Catalytic Combustion of Natural Gas

Catalytic combustion is a cleaner and more efficient method of burning natural gas compared to traditional flame combustion. It involves using a catalyst, typically a metal oxide, to promote the oxidation of methane (the primary component of natural gas) at lower temperatures, reducing the formation of pollutants like NOx and CO. REEs play a pivotal role as catalysts in the combustion of natural gas, contributing to cleaner and more efficient energy conversion. Their oxides, particularly those of Ce, La, and Pr are widely exploited due to their excellent properties, thermal stability, and oxygen storage-release capacity. CeO_2, for example, can rapidly cycle between Ce^{4+} and Ce^{3+} states, facilitating the oxidation of methane and other hydrocarbons at lower temperatures. This not only improves combustion efficiency but also helps reduce harmful emissions of CO and unburned hydrocarbons.

In an elegant work, a supramolecular strategy was employed to address the limitations of methane-oxidation catalysts. Specifically, single nanostructures consisting of a Pd core and a CeO_2 shell were preassembled in solution and uniformly deposited onto a modified hydrophobic alumina support (Cargnello et al., 2012). Structural characterization, including electron microscopy, confirmed that the Pd cores remained well-dispersed and isolated even after exposure to high temperatures up to 850°C. This nanoarchitecture resulted in complete conversion of methane at relatively low temperature. Similarly, Peng et al. (2018) developed thermally stable Pd-CeO_x nanowires having porous silica shells for complete oxidation of methane at 350°C, showcasing the efficacy of REEs in the combustion of methane.

3.5 REEs as Catalysts for Gases and Volatile Organic Compounds

REEs play a vital role as purification catalysts in the treatment of industrial waste air by facilitating the removal of harmful pollutants such as volatile organic compounds (VOCs), NO_x, and CO. Their unique redox properties, especially when incorporated into metal oxides like CeO_2, enhance oxygen storage and release, enabling efficient oxidation reactions even at lower temperatures (Hu et al., 2019; Zhan et al., 2014). A nanoporous NiCuMnO catalyst exhibited high activity and stability for NO reduction and CO oxidation (Fujita et al., 2016). Microstructural analysis revealed that active Cu/CuO regions were intertwined with a robust NiMnO nanoporous framework. *In situ* TEM during NO reduction captured a distinctive reaction-induced self-transformation of the nanostructure, highlighting its dynamic catalytic behaviour. Recently, a mesoporous Pt/CeO_2 (MOF) catalyst was derived from Pt–CeBDC via high-temperature treatment for efficient low-temperature HCHO oxidation (Shi et al., 2023). Compared to a commercial Pt/CeO_2 catalyst (31% conversion), Pt/CeO_2 (MOF) achieved 78% HCHO conversion at room temperature with excellent long-term stability. The enhanced activity is attributed to highly dispersed Pt nanoparticles within CeO_2 mesopores, generating abundant Pt-O-Ce interfaces and ROS (e.g., O_2^{2-}). Kinetic and DRIFTS analyses confirmed accelerated HCHO decomposition and CO oxidation steps, highlighting the superior catalytic performance of the material.

3.6 REEs as Catalysts in Typical Organic Syntheses

RE metals are emerging as powerful yet underexplored tools in modern synthetic chemistry. While transition metal-mediated transformations, driven by well-established mechanisms such as oxidative addition and reductive elimination, have long dominated the field, RE metals offer a distinct reactivity profile, facilitating key nucleophilic substitution, conjugate addition, bond insertions, and carbon-

carbon (C-C) coupling reactions. All this progress underscores the versatility of RE metal catalysts, holding great promise in contemporary organic synthesis (Sun et al., 2024). Many REE salts, particularly lanthanide triflates [$Ln(OTf)_3$], are strong Lewis acids that can efficiently such as $La(OTf)_3$ and $Yb(OTf)_3$ catalyze aldol condensations under mild, water-tolerant conditions (Kobayashi, S. & Hachiya, I., 1994). Their high oxophilicity makes them particularly suitable for reactions involving oxygen- or nitrogen-containing substrates.

C–C coupling reactions are pivotal in constructing complex organic molecules by forming new C–C bonds between two carbon-containing units. These reactions, often catalysed by metals, are widely employed in pharmaceuticals, agrochemicals, and material synthesis. Their efficiency, selectivity, and broad substrate scope make them essential tools in modern organic chemistry (Garg, Bisht, & Ling, 2016; Garg & Bisht 2025). Alongside, MOFs based on REEs combine the high coordination versatility of RE metals with organic linkers to form highly porous, thermally stable structures, making them valuable for catalysis applications. For instance, A holmium-based MOF, [$Ho(2\text{-}TriPP\text{-}COO)_3$], was synthesized via a hydrothermal method using terpyridine-functionalized ligands (Dong et al., 2018). Its structure was confirmed through single-crystal and powder XRD along with elemental analysis. Leveraging the strong affinity between Pd^{2+} ions and uncoordinated polypyridine sites, a post synthetic Pd incorporation yielded a bifunctional Pd-HoMOF catalyst. This catalyst showcased remarkable efficiency in Heck and Suzuki–Miyaura cross-coupling reactions, combining the structural robustness of MOFs (stable up to 400 °C) with the reactivity of homogeneous Pd catalysts. Furthermore, it demonstrated excellent recyclability over multiple cycles, highlighting its potential for sustainable fine chemical synthesis. Current research trends reflect a strong shift toward employing REE-MOFs at the interface of organic synthesis and sustainable catalysis, offering promising platforms for green and efficient chemical transformations, including, but not limited to, Michael addition, Knoevenagel condensation, cyanosilylation, cycloaddition, oxidation, and reduction reactions.

4. Conclusions

The strategic deployment of REEs in catalysis has significantly transformed the landscape of sustainable chemical processes. Their unique 4f-electronic configurations, variable oxidation states, and high coordination adaptability offer unmatched opportunities to design robust and selective catalytic systems. From wastewater treatment to petroleum refining, and from NH_3 synthesis to fine organic transformations, REE-based catalysts are proving indispensable across a broad spectrum of applications, addressing both environmental and industrial demands. In this context, the marriage of REEs with nanomaterials and MOFs is acting as a bridge between homo- and heterogeneous catalysis, influencing the current sustainable chemistry. Needless to say, REEs are here to stay to build a relatively more robust and sustainable chemical industry.

Acknowledgements

B. Garg acknowledges the Research & Development Cell of IIMT University, Meerut, India. T. Bisht is thankful to the Department of Science & Technology (DST), New Delhi, India for the Curie Grant (DST/CURIE-PG/2022/90(C)).

References

Ahmadi, S., Rahdar, A., Igwegbe, C.A., Mortazavi-Derazkola, S., Banach, A.M., Rahdar, S., Singh, A.K., Rodriguez-Couto, S. and Kyzas, G.Z. (2020). Praseodymium-doped cadmium tungstate ($CdWO_4$) nanoparticles for dye degradation with sonocatalytic process. *Polyhedron*, 190, 114792.

Akah, A. (2024). Role of Rare Earths as Catalysts in the Chemical, Petroleum, and Transportation Industries. In: Murty, Y.V., Alvin, M.A., Lifton, J. (Eds.), *Rare Earth Metals and Minerals Industries*. Springer, Cham. https://doi.org/10.1007/978-3-031-31867-2_13.

Anastas, Paul T. and John C. Warner. (2000). *Green Chemistry: Theory and Practice* (Oxford, 2000; online Edn., Oxford Academic, 31 Oct. 2023), https://doi.org/10.1093/oso/9780198506980.001.0001, accessed 24 June 2025.

Asadollahzadeh, M., Torkaman, R. and Mostaedi, M.T. (2020). Extraction and Separation of Rare Earth Elements by Adsorption Approaches: Current Status and Future Trends. *Sep. Purif. Rev.*, *50*, 417–444.

Bünzli, J.-C.G. (2015). On the design of highly luminescent lanthanide complexes. *Coord. Chem. Rev.*, 293–294, 19–47.

Bünzli J.-C. G. and Pecharsky V.K. (Eds.) 2013. *Handbook on the Physics and Chemistry of Rare Earths* (Vol. 44). North Holland. ISBN 978-0444627117.

Cao, R., Huang, H., Tian, N., Zhang, Y., Guo, Y. and Zhang, T. (2015). Novel Y doped Bi_2WO_6 photocatalyst: Hydrothermal fabrication, characterization, and enhanced visible-light-driven photocatalytic activity for Rhodamine B degradation and photocurrent generation. *Mater. Charact.*, 101, 166–172.

Cargnello, M., Jaén, J.J.D., Garrido, J.C.H., Bakhmutsky, K., Montini, T., Gámez, J.J.C., Gorte, R.J. and Fornasiero, P. (2012). Exceptional Activity for Methane Combustion over Modular Pd@ CeO_2 Subunits on Functionalized Al_2O_3. *Science*, 337, 713–717.

Chen, X., Zhang, M., Qin, H., Zhou, J., Shen, Q., Wang, K., Chen, W., Liu, M. and Li, N. (2022). Synergy effect between adsorption and heterogeneous photo-Fenton-like catalysis on $LaFeO_3$/lignin-biochar composites for high efficiency degradation of ofloxacin under visible light. *Sep. Purif. Technol.*, 280, 119751.

Cherkezova-Zheleva, Z., Burada, M., Sobetkii, A.E., Paneva, D., Fironda, S.A. and Piticescu, R.-R. (2024). Green and sustainable rare earth element recycling and reuse from end-of-life permanent magnets. *Metals*, 14, 658.

Cymes, B.A., Almquist, C.B. and Krekeler, M.P. (2020). Europium-doped cryptomelane: Multi-pathway synthesis, characterization, and evaluation for the gas phase catalytic oxidation of ethanol. *Appl. Catal. A Gen.*, 589, 117310.

Dagwar, P.P., Iqbal, S.S. and Dutta, D. (2025). Sustainable recovery of rare earth elements from industrial waste: A path to circular economy and environmental health. *Waste Manag. Bull.*, 3, 373–390.

Dong, D., Li, Z., Liu, D., Yu, N., Zhao, H., Chen, H., Liu, J. and Liu, D. (2018). Postsynthetic modification of single Pd sites into uncoordinated polypyridine groups of an MOF as the highly efficient catalyst for Heck and Suzuki reactions. *New J. Chem.*, 42, 9317–9323.

El-Kholy, R.A., Isawi, H., Zaghlool, E., Soliman, E.A., Khalil, M.M.H., Said, M.M. and El-Aassar, A.-E. (2023). Preparation and characterization of rare earth element nanoparticles for enhanced photocatalytic degradation. *Environ. Sci. Pollut. Res.*, 30, 69514–69532.

Ertl, G. (2008). Reactions at Surfaces: From Atoms to Complexity (Nobel Lecture). *Angew. Chem. Int. Ed.*, 47, 3524–3535.

Fan, Z., Wu, T. and Wu, X. (2017). Synthesis of reduced graphene oxide as a platform for loading β-$NaYF_4$:Ho^{3+}@TiO_2 based on an advanced visible light-driven photocatalyst. *Sci. Rep.*, 7, 13833.

Fida, H., Zhang, G., Guo, S. and Naeem, A. (2017). Heterogeneous Fenton degradation of organic dyes in batch and fixed bed using La-Fe montmorillonite as catalyst. *J. Colloid Interf. Sci.*, 490, 859–868.

Fouad, K., Alalm, M.G., Bassyouni, M. and Saleh, M.Y. (2021). Optimization of catalytic wet peroxide oxidation of carbofuran by Ti-$LaFeO_3$ dual photocatalyst. *Environ. Technol. Innov.*, 23, 101778.

Fujita, T., Abe, H., Tanabe, T., Ito, Y., Tokunaga, T., Arai, S., Yamamoto, Y., Hirata, A. and Chen, M. (2016). Earth-Abundant and Durable Nanoporous Catalyst for Exhaust-Gas Conversion. *Adv. Funct. Mater.*, 26, 1609–1616.

Garg, B., Bisht, T. and Thomas, K.R.J. (2017). Magnetic graphene nanocomposites for multi-functional applications. In: . Prof. S.K. Sharma (Ed.) *Complex Magnetic Nanostructures: Synthesis, Assembly, and Applications*. Springer International Publishing AG, Cham, Chapter 9, pp. 317–357.

Garg, B. and Ling, Y.-C. (2013). Versatilities of graphene-based catalysts in organic transformations. *Green Mater.*, 1, 47–61.

Garg, B., Bisht, T.M. and Ling, Y.-C. (2016). Graphene-based nanomaterials: Versatile catalysts for carbon-carbon bond forming reactions. *Curr. Org. Chem.*, 20, 1547–1566.

Garg, B. and Bisht, T. (2016). Carbon nanodots as peroxidase nanozymes for biosensing. *Molecules*, 21, 1653.

Garg, B., Bisht, T. and Ling, Y.-C. (2015). Graphene-based nanomaterials as efficient peroxidase mimetic catalysts for biosensing applications: An overview. *Molecules*, 20, 14155–14190.

Garg, B., Bisht, T. and Ling, Y.-C. (2014a). Graphene-based nanomaterials as heterogeneous acid catalysts: A comprehensive perspective. *Molecules*, 19, 14582–14614.

Garg, B., Bisht, T. and Ling, Y.-C. (2014b). Sulfonated graphene as highly efficient and reusable acid carbocatalyst for the synthesis of ester plasticizers. *RSC Adv.*, 4, 57297–57307.

Garg, B. and Ling, Y.-C. (2018). Carbon-based nanomaterials as nanozymes. In: Prof. Klaus D. Sattler (Ed.), *Carbon Nanomaterials Sourcebook, Nanoparticles, Nanocapsules, Nanofibers, Nanoporous Structures, and Nanocomposites* (Vol. II). Taylor & Francis, CRC Press, Chapter 14, pp. 309–333.

Garg, B. (2018). ^{31}P solid-state NMR spectroscopy of adsorbed phosphorous probe molecules: Acidity characterization of solid acid carbonaceous materials for catalytic applications. In: S.K. Sharma (Ed.), *Handbook of Materials Characterization*. Springer International Publishing AG, Cham, Chapter 14, pp. 549–596.

Garg, B. and Bisht, T. (2025). Nanocomposites as catalysts in organic coupling reactions. In: S.K. Swain (Ed.), Nanocomposites as catalysts in organic reactions. *Royal Society of Chemistry, Chapter 12*, pp. 337-368. Doi: https://doi.org/10.1039/9781837677191-00337

Gogoi, A., Navgire, M., Sarma, K.C. and Gogoi, P. (2017). Fe_3O_4-CeO_2 metal oxide nanocomposite as a Fenton-like heterogeneous catalyst for degradation of catechol. *Chem. Eng. J.*, 311, 153–162.

Ho, C.-C., Kang, F., Chang, G.-M., You, S.-J. and Wang, Y.-F. (2019). Application of recycled lanthanum-doped TiO2 immobilized on commercial air filter for visible-light photocatalytic degradation of acetone and NO. *Appl. Surf. Sci.*, 465, 31–40.

Hu, Q., Cao, K., Lang, Y., Chen, R., Chu, S.Q., Jia, L.W., Yue, J. and Shan, B. (2019). Improved NO-CO reactivity of highly dispersed Pt particles on CeO_2 nanorod catalysts prepared by atomic layer deposition. *Catal. Sci. Technol.*, 9, 2664–2672.

Karunakaran, C., Gomathisankar, P. and Manikandan, G. (2010). Preparation and characterization of antimicrobial Ce-doped ZnO nanoparticles for photocatalytic detoxification of cyanide. *Mater. Chem. Phys.*, 123, 585–594.

Kobayashi, S. and Hachiya, I. (1994). Lanthanide triflates as water-tolerant Lewis acids. Activation of commercial formaldehyde solution and use in the aldol reaction of silyl enol ethers with aldehydes in aqueous media. *J. Org. Chem.*, 59, 3590–3596.

Kousar, T., Aadil, M. Zulfiqar, S., Ibrahim, S.M., Mubeen, S., Hassan, W., Shafiq, K. and Mahmood, F. (2024). Facile synthesis of graphene anchored rare earth doped mixed metal ferrite nanorods: A potential candidate for azo dye mineralization. *J. Rare Earths*, 42, 907–916.

Li, L., Zhang, T., Zhou, Y., Wang, X., Au, C.-t. and Jiang, L. (2022). Review on catalytic roles of rare earth elements in ammonia synthesis: Development and perspective. *J. Rare Earths*, 40, 1–10.

Liang, Z., Yin, L., Yin, H., Yin, Z. and Du, Y. (2022). Rare earth element based single-atom catalysts: synthesis, characterization and applications in photo/electro-catalytic reactions. *Nanoscale Horiz.*, 7, 31–40.

Luan, J., Liu, W., Yao, Y., Ma, B., Niu, B., Yang, G. and Wei, Z. (2022). Synthesis and property examination of $Er_2FeSbO_7/BiTiSbO_6$ heterojunction composite catalyst and light-catalyzed retrogradation of enrofloxacin in pharmaceutical waste water under visible light irradiation. *Materials*, 15, 5906.

Murata, S. and Aika, K.I. (1992). Preparation and characterization of chlorine-free ruthenium catalysts and the promoter effect in ammonia synthesis: 2. A lanthanide oxide-promoted Ru/Al_2O_3 catalyst. *J. Catal.*, 136, 118–125.

Niwa, Y. and Aika, K.I. (1996). Ruthenium catalyst supported on CeO_2 for ammonia synthesis. *Chem. Lett.*, 25, 3–4.

Niwa, Y. and Aika, K.I. (1996). The effect of lanthanide oxides as a support for ruthenium catalysts in ammonia synthesis. *J. Catal.*, 162, 138–142.

Ogura, Y., Sato, K., Miyahara, S.-I., Kawano, Y., Toriyama, T., Yamamoto, T., Matsumura, S., Hosokawa, S. and Nagaoka, K. (2018). Efficient ammonia synthesis over a $Ru/La_{0.5}Ce_{0.5}O_{1.75}$ catalyst pre-reduced at high temperature. *Chem Sci.*, 9, 2230–2237.

Ortúzar, M., Esterhuizen, M., Olicón-Hernández, D.R., González-López, J. and Aranda, E. (2022). Pharmaceutical Pollution in Aquatic Environments: A Concise Review. *Front. Microbiol.*, 13, 869332.

Owens, C.L., Nash, G.R., Hadler, K., Fitzpatrick, R.S., Anderson, C.G., and Wall, F. (2019). Apatite enrichment by rare earth elements: A review of the effects of surface properties, *Adv. Colloid Interface Sci.*, 265, 14–28.

Özdemir, H., Çiftçioğlu, E. and Öksüzömer, M.A.F. (2023). Lanthanum based catalysts for oxidative coupling of methane: Effect of morphology and structure. *Chem. Eng. Sci.*, 270, 118520.

Patel, D.K., Sharma, V.K., Kumar, V., Kumar, P. and Kumar, D. (2025). Enhancing hydrophobicity and anti-corrosion properties of Al-6061Aluminum hybrid composites through trace additions of rare earth oxides. In: *Proceedings of the Institution of Mechanical Engineers, Part E: Journal of Process Mechanical Engineering*, 09544089251317381.

Patel, D.K., Sharma, V.K. and Sharma, H.O. (2025). Current global Status of rare earth elements (REEs) and their role as catalysts in reducing air pollution for maintaining environmental sustainability. *Transactions of the Indian Institute of Metals*, 78(1), 4.

Peng, H., Rao, C., Zhang, N., Wang, X., Liu, W., Mao, W., Han, L., Zhang, P. and Dai, S. (2018). Confined Ultrathin Pd-Ce Nanowires with Outstanding Moisture and SO_2 Tolerance in Methane Combustion. *Angew Chem. Int. Ed. Eng.*, 57, 8953–8957.

Poliakoff, M., Fitzpatrick, J.M., Farren, T.R. and Anastas, P.T. (2002). Green Chemistry: Science and Politics of Change. *Science*, 297, 807-810.

Rosid, S.J.M., Toemen, S., Iqbal, M.M.A., Wan, W.A., Bakar, A., Mokhtar, W.N.A.W. and Aziz, M.M. A. (2019). Overview performance of lanthanide oxide catalysts in methanation reaction for natural gas production. *Environ. Sci. Pollut. Res.*, 26, 36124–36140.

Seroglazova, A.S., Chebanenko, M.I., Nevedomskyi, V.N. and Popkov, V.I. (2023). Solution combustion synthesis of novel $PrFeO_3/CeO_2$ nanocomposite with enhanced photo-Fenton activity under visible light. *Ceram. Int.*, 49, 15468–15479.

Shannon, R.D. (1976). Revised effective ionic radii and systematic studies of interatomic distances in halides and chalcogenides. *Acta Crystallographica Section A*, 32, 751–767.

Shao, Q., Wang, P., Tian, H., Yao, R., Sun, Y. and Long, J. (2009). Study of the application of structural catalyst in naphtha cracking process for propylene production. *Catal. Today*, 147, S347–S351.

Sheldon, R.A. (2018). Metrics of Green Chemistry and Sustainability: Past, Present, and Future. *ACS Sustain. Chem. Eng.*, 6, 32-48.

Shi, K., Lv, S., Hua, Z., Ruan, C., Liu, X., Wu, N., Mao, C. and Li, L. (2023). Highly active low-temperature HCHO oxidation of mesoporous Pt/CeO_2 derived from Pt/CeBDC. *New J. Chem.*, 47, 21905–21915.

Sin, J.-C., Lam, S.-M., Lee, K.-T. and Mohamed, A.R. (2013). Preparation and photocatalytic properties of visible light-driven samarium-doped ZnO nanorods. *Ceram. Int.*, 39, 5833–5843.

Singh, L., Rekha, P. and Chand, S. (2016). Cu-impregnated zeolite Y as highly active and stable heterogeneous Fenton-like catalyst for degradation of Congo red dye. *Sep. Purif. Technol.*, 170, 321–336.

Sun, D., Xu, J., Liu, G. and Chen, Y.-H. (2024). Rare Earth Metal Reagents in Organic Synthesis: A Comprehensive Review. *Eur. J. Org. Chem.*, 28, e202400540.

Sundararaman, S., Kumar, J.A., Deivasigamani, P. and Devarajan, Y. (2022). Emerging pharma residue contaminants: Occurrence, monitoring, risk, and fate assessment: A challenge to water resource management. *Sci. Total Environ.*, 825, 153897.

Suresh, M., Radhakrishan, P. and Sivasamy, A. (2024). Solar driven highly efficient photocatalyst based on Dy_2O_3 nanorods deposited on reduced graphene oxide nanocomposite for methylene blue dye degradation. *Environ. Sci. Pollut. Res.*, 31, 60260–60278.

Tunsu, C., Petranikova, M., Gergorić, M., Ekberg, C. and Retegan, T. (2015). Reclaiming rare earth elements from end-of-life products: A review of the perspectives for urban mining using hydrometallurgical unit operations. *Hydrometallurgy*, 156, 239–258.

Xu, L. and Wang, J. (2012). Magnetic nanoscaled Fe_3O_4/CeO_2 composite as an efficient Fenton-like heterogeneous catalyst for degradation of 4-chlorophenol. *Environ. Sci. Technol.*, 46, 10145–10153.

Younes, H.A., Taha, M., Khaled, R., Mahmoud, H.M. and Abdelhameed, R.M. (2023). Perovskite/metal-organic framework photocatalyst: A novel nominee for eco-friendly uptake of pharmaceuticals from wastewater. *J. Alloys Compd.*, 930, 167322.

Zhou, C., Zhu, Y. and Liu, H. (2010). Effect of samarium on methanation resistance of activated carbon supported ruthenium catalyst for ammonia synthesis. *J. Rare Earths*, 28, 552–555.

Zhu, Y.F., Li, X.N., Zhou, C.H., Gao, D.M., & Liu, H.Z. (2003). Study on chemisorption and desorption of hydrogen and nitrogen on Ru-based ammonia synthesis catalyst. *Chin. J. Chem. Eng.*, 11, 15–18.

5 Sustainable Transportation and Rare Earth Elements: Consumer Perception and Purchase Intentions for Electric Vehicles

Ashish Kumar[1*], Puneet Kumar[2] and Devika Agarwal[3]

1. Introduction

Fossil fuels are regarded as the principal energy source in the current global context. With the rise of urbanization and the migration of populations to urban areas, our total energy consumption continues to escalate. Regrettably, the combustion of fossil fuels generates substantial greenhouse gas (GHG) emissions, which are the principal cause of climate change and associated issues (Gahlaut et al., 2024; Kamran et al., 2021). The transportation industry now accounts for around 23% of global greenhouse gas emissions, a figure projected to escalate to 50% by 2030, as per the International Energy Agency (Jayaswal et al., 2022). This considerable increase suggests that our existing transportation system is unsustainable and requires substantial modifications to mitigate its environmental effects (Egbue & Long, 2012). The primary determinants driving the expansion of the automotive sector will be urbanization and the resulting increase in vehicle use, particularly in emerging economies like India.

Electric vehicles (EVs) are growing as an alternative green technology that may address these challenges and facilitate the effective transition to a sustainable, low-carbon transportation system while conserving limited natural resources (Tarei et al., 2021; Patel et al., 2024). Consumers are important in realizing the sustainability advantages of technology breakthroughs via their acceptance and adoption. Policymakers' strategy actions significantly influence consumer sentiments about these shifts (Ozaki & Sevastyanova, 2011). The Government of India recognizes the significance of EVs in establishing a sustainable transportation system and is actively developing laws and incentives to encourage their adoption

[1] Assistant Professor, Faculty of Management & Commerce, Poornima University, Jaipur, Rajasthan, India.
[2] Associate Professor, School of Management and Commerce, IIMT University, Meerut Uttar Pradesh, India.
[3] Professor, Faculty of Management & Commerce, Poornima University, Jaipur, Rajasthan, India.
* Corresponding author: ashishdeshwal49@gmail.com

DOI: 10.1201/9781003683254-5

(Digalwar & Rastogi, 2023). The Indian government seeks a substantial transition to electric transportation, committing to a 33–35% reduction in its carbon footprint by 2030 compared to 2005 levels. India has established a national objective of attaining 30% EV market penetration by 2030 and initiated a National Mission on Transformative Mobility and Battery Storage to encourage the localization of EV component manufacture.

In accordance with central government measures, several state governments have implemented policies to promote EVs (Rocky Mountain Institute & Niti Aayog, 2019). These policies have resulted in diverse incentives, infrastructure enhancement, financial allocation, research and development, production, and sales (Das & Bhat, 2022). Despite governmental initiatives, India's adoption rate of EVs remains below projections and the worldwide average (Kumar et al., 2020). Consequently, examining customer perception across several geographic areas of the nation is essential for addressing the EV disparity. The rapid increase in vehicles, especially on roadways, intensifies the issue. The problem is exacerbated by the concerning degree of air pollution in Indian cities. A staggering 64 Indian cities are among the top 100 most polluted cities worldwide (IQAir, 2024). India has a critical challenge in reconciling economic expansion, driven by advancements in the transportation sector, with environmental sustainability. Confronting the issue of rising carbon emissions from the transportation sector is essential for enhancing air quality and securing a healthy future for the populace. This study tries to analyse consumer views and intention to adopt EVs within the framework of a rising economy. The research focuses on consumers' viewpoints in India; nevertheless, we will use a structural equation modelling approach to identify the variations in perceptions across several places within India's rising economy.

1.1 Sustainable Transportation and Rare Earth Elements

The global transportation sector is being transformed more deeply due to an increasingly green-conscious public, rapid advances in technology, and an urgent need to reduce GHG emissions. Among the most encouraging new trends are electric-power vehicles (EVs). They are beginning to be seen as an indispensable link for sustainable development. Unlike traditional internal combustion engine vehicles, EVs offer the potential for emissions to be much lower over their lifetime. However, their production depends very heavily on Rare Earth Elements (REEs) (Sharma et al., 2019; Patel et al., 2025; Srikant et al., 2021). Materials like lithium, cobalt, and neodymium carry their own separate set of environmental and ethical dilemmas outwith the automotive sector altogether. As the rate of EV adoption rises, consumer perception becomes essential to shaping market trends and policy effectiveness. Even though many people are enthusiastic about the environment-friendly aspects of EVs, worries about resource extraction, battery life, charging facilities, and overall sustainability continue to nag them. Understanding how consumers weigh these trade-offs and how such evaluations sway their purchasing intentions is crucial for governments, manufacturers, and pressure groups trying out the new transportation models.

Furthermore, the objectives of the study are to examine how Indian consumers incorporate sustainability factors into their EV purchase choices and how these concerns vary across diverse populations. Consequently, the study findings might assist policymakers and automobile manufacturers in re-evaluating and modifying their plans to enhance sustainability within the Indian transportation industry. In accordance with our intention, we formulated the following objectives of the research.

1.2 Objectives

1. To identify the key factors influencing consumer decision-making regarding the purchase of EVs.
2. Analyse the role of environmental consciousness in shaping purchase intentions for EVs.
3. Assess the impact of government policies and incentives on consumer willingness to adopt EVS

1.3 Review of Literature and Hypothesis Development

1.3.1 Worldwide Context of EV Development

The world's dominant energy source, considered from a global perspective, is fossil fuel. Our overall consumption of energy is constantly increasing with the increasing urbanization and migration of people towards cities. Unfortunately, burning of fossil fuels results in a considerably high GHG emission, which is the primary contributor to climate change and related problems (G. Singh et al., 2024). Recognizing the importance of EV adoption, governments throughout the world have implemented a variety of financial and non-financial incentives to promote EV usage. Economic incentives such as tax breaks, rebates, and exemptions from registration fees and taxes (ACEA, 2024; Ghatikar et al., 2017).

Furthermore, several countries provide non-monetary incentives, such as access to high-occupancy vehicle (HOV) lanes and free parking (Ghatikar et al., 2017). As a result, the EV sector has grown steadily in recent years, and the global inventory will approach 10 million units in 2020, a significant 43% increase over 2019. Battery electric vehicles (BEVs) accounted for 67% of all EV registrations in 2020 (Williams, 2023). With 4.5 million EVs on the road, China has emerged as the market leader. Europe follows closely after with the most annual registrations, reaching an astonishing 3.2 million units. Notably, Norway has the highest per capita ownership of electric cars worldwide. In 2020, Europe led the way in new electric car registrations, accounting for 1.4 million vehicles. China registered 1.2 million electric cars, while the United States recorded 295,000. Germany and France experienced significant increases, with 395,000 and 185,000 new registrations, respectively. In 2020, the UK registered 176,000 new EVs (Williams,

2023). Several European countries, including Germany, the United Kingdom, France, Norway, Italy, Switzerland, the Netherlands, and Sweden, have emerged as key EV markets. Furthermore, nations such as Australia, Austria, Canada, Spain, Italy, New Zealand, and Japan are seeing significant growth in EV use. Norway demonstrated its status as an EV pioneer by reaching a record sales share of 75% in 2020 compared to the previous year (IEA, 2023). Similarly, Ireland, Sweden, and the Netherlands had significant sales increases of 50%, 30%, and 25%, respectively, highlighting the growing popularity of electric cars in these markets (Williams, 2023). The fast advancement of EV technology is driving a surge in electric two- and three-wheelers, buses, and trucks (Electric Vehicle Industry Report, 2024). This trend suggests that electric mobility will become more common in a variety of transportation sectors.

1.3.2 EV Context from India

Emerging economies, such as India, are struggling with the issue of environmental degradation. They are gradually favouring EVs in response to the global move toward electric mobility (Wu & Zhang, 2017). In 2012, the Indian government announced the National Electricity Mobility Mission Plan (NEMMP), a comprehensive national policy programme that signified a big step forward in EV marketing. The FAME-India initiative was introduced in 2015 as part of the National Electric Mobility Mission Plan (NEMMP). Phase I of the FAME-India initiative was first proposed as a two-year endeavour commencing on April 1, 2015, but was then extended until March 31, 2019. The first phase of FAME-India, with an INR 895 crore budget, concentrated on four areas to strengthen the country's EV industry: technological development, demand generation, charging infrastructure, and pilot projects (Kohli, 2024). A government assessment of FAME I's accomplishments found that fuel conservation and CO2 emission reductions did not meet the expectations, the market reacted slowly and hesitantly, the subsidy framework needed to be reformed to establish equity among emerging EV technologies, and the electric three-wheeler sector expanded unexpectedly quickly (EVINDIA, 2024). The Government of India launched FAME II in April 2019, with a USD 1.4 billion budget, to incorporate these inputs and insights. This intends to promote widespread use of EVs and charging infrastructure, with a focus on megacities, national roadways, and corridors. The goal is to build a strong domestic electric car ecosystem. FAME II was slated to run from 2019 to 2022, with a substantially larger budget than FAME I at INR 100 billion (about USD 1.45 billion). FAME II's duration has been extended by two years, and it will now finish in March 2024 (Gulati, 2022). In addition to federal government initiatives, numerous Indian state governments have passed legislation to stimulate the use of EVs within their respective jurisdictions. These policies aim to make their states desirable hubs for electric car production, assembly, sales, and servicing (Gupta et al., 2020). Standard components of state-level electric car policies include initial capital subsidies for EV purchases, exemptions from road taxes and registration fees, and financial aid for the development of charging infrastructure and services.

1.3.3 Customers' Perspectives and Purchasing Intentions for EVs

Several elements of EVs have significant effects on consumer purchase decisions (Mandys, 2021). Customers' purchase decisions are related to specifications for different electric car types. Customers in the pre-purchase phase put a high value on variables including range, battery capacity, charge time, acceleration, peak speed, and price value. These variables are associated with the purchase intentions that consumers consider. Frequent electric car users are aware of these characteristics since they are more aware of the considerations required when purchasing a new EV model. The adoption of electric cars in Southeast Asian countries is still in its early stages, and more time is needed to assess sustainable vehicle models for everyday usage thoroughly.

Furthermore, customers' purchasing intentions are impacted mainly by their financial capability, since the price range for EVs varies depending on the model and its features. The majority of the audience may struggle to purchase expensive EV models owing to the initial financial expenditure and concerns about maintenance expenses (Slowik et al., 2022). As a result, a variety of factors impact customers' purchasing decisions. The adoption of EVs is linked to the positive driving experience that customers will have while using them. Furthermore, buyers' valuations neglect the whole distance an EV can go on a single charge (Simolin, 2022).

Furthermore, they assess the battery's energy retention performance, including its charging time. Additional parameters of EVs are also evaluated, such as acceleration rate and maximum achievable speed. Finally, the price margin of the car model being considered during the pre-purchase stage has a significant impact on consumers' buying choices. Approximately 70% of tier-one people favour electric cars as their future form of transportation, demonstrating widespread adoption of digital EV models in India (Chanda et al., 2024). Customers' buying behaviour suggests that they are prepared to pay a premium for electric automobiles based on their relevant characteristics and perceived ease of usage (Featherman et al., 2021). Electric cars may be too costly for many Indians; nonetheless, their increasing acceptance suggests a greater willingness among the Indian population to adopt EVs over conventional ones.

1.3.4 Hypothesis Development

Cost savings factors: One of the key motivators for people considering EVs is the prospect of long-term financial savings. In contrast to usual internal combustion engine cars, EVs generally have lower operating costs due to lower fuel prices, fewer mechanical components that need maintenance, and different government subsidies or tax breaks granted in many countries (Alanazi, 2023). These financial advantages are particularly appealing to cost-conscious buyers who base their automotive purchasing choices on the overall cost of ownership over time. Previous research has shown that economic factors, notably predicted fuel and maintenance savings, have a significant influence on EV adoption (K V et al., 2022; Phuthong et al., 2024; Vafaei-Zadeh et al., 2025). Consumers see EVs as a realistic alternative not just for environmental sustainability, but also for personal economic efficiency due to lower long-term driving expenses (Buhmann & Criado, 2023). With increased knowledge and accessibility to these vehicles, the reported price benefits are becoming a significant factor in changing customer choices for alternative fuel alternatives. Therefore, it is hypothesized that:

H1: Cost savings factors positively influence consumers' intention to purchase an electric car.

Technological advancements factors: Technology has a significant role in shaping customer preferences, particularly for sustainable mobility options like electric and flex-fuel cars. Advancements in fuel efficiency, engine performance, hybrid powertrains, and infrastructure (such as charging stations and ethanol availability) have led to growing adoption of alternative fuel cars as a realistic and value-added alternative to conventional vehicles (Delso-Vicente et al., 2025; Velho et al., 2024). Advancements in technology, such as safety features, on board diagnostics, and intuitive digital interfaces, enhance the user experience and promote customer confidence. As technology advances, buyers may see flex-fuel cars as a reliable and future-proof alternative to regular fossil fuel automobiles (Yalavarthy et al., 2025). Previous research has shown that when customers see innovative, efficient, and easy technology, their inclination to acquire alternative fuel cars improves dramatically (Kowalska-Pyzalska et al., 2022). Therefore, in the context of this study, it is hypothesized that:

H2: Technological advancements positively influence consumers' intention to purchase an electric car.

Sustainability concerns: The worldwide trend toward more sustainable practices has sparked renewed interest in ecologically friendly types of transportation, notably electric cars (Tilly et al., 2024). Consumers today are becoming more aware of the negative repercussions of fossil fuel consumption, such as GHG emissions that pollute the air we breathe and deplete resources critical to our existence (Talukdar et al., 2024). As a result, concern for environmental preservation has evolved as a powerful psychological motivation for consumer behaviour. Individuals who are environmentally concerned are more likely to choose sustainable options such as electric cars over gasoline-powered vehicles. Previous scientific work has shown that environmental sustainability has an essential role in developing 'green' buying intentions and is directly related to the adoption of low-emission technologies (Gulzari et al., 2022). Given these conditions, it stands to reason that as consumers become more aware of the environmental concerns associated with conventional fuels, their desire to switch to electric cars increases. Therefore, it is hypothesized that:

H3: Sustainability concerns positively influence consumers' intention to purchase an electric car.

Government policies: The government plays an essential role in changing consumer shopping behaviour, and it is also the most effective way to promote low-carbon transportation. Policies such as subsidies and tax breaks for EVs, incentives for manufacturers, environmental restrictions, and long-term energy targets have an impact on how the public will perceive and adopt clean fuel technology (Kresnanto & Putri, 2024). Various countries have adopted targeted policies for the promotion of electric and flexible-fuel vehicles, such as infrastructure development, ethanol blending laws, and economic incentives (Malik et al., 2025). Researchers believe that interventions may help decrease perceived costs and risks of new auto technology, making them more appealing to customers. In India, schemes like "Speedier Adoption and Production of Hybrid and Electric Vehicles" influenced consumer awareness and buying behaviour to a great extent (Nimesh et al., 2024). Considering the importance of government factors to promote green mobility, we can thus formulate a hypothesis:

H4: Government policies positively influence consumers' intention to purchase an electric car.

Social factors: Among these causes are conventions and the impact of peer groups, cultural beliefs, and lifestyle trends, as well as a growing awareness that sustainability and climate change are serious public concerns (Goggins et al., 2022). As electric cars become more popular, customers are more likely to react to remarks from family members, friends, and community organizations (Tu & Yang, 2019). However, reviews are difficult. Previous studies have shown that social variables impact people's intentions to utilize green technologies. If consumers witness friends or others in positions of prominence purchasing electric cars, they will believe the EV is superior or at least worth considering as part of their auto fleet. In urban and middle-class culture, owning an electric automobile is a statement of social duty and environmentally conscious beliefs (Electric Vehicle Industry Report, 2025). People's moral and ethical standards are shifting as environmental notions become more widely communicated via social media, public service initiatives, and education programmes. In fact, these patterns imply that social considerations may play a key role in consumer EV buying decisions. Therefore, it is hypothesized that:

H5: Social factors positively influence consumers' intention to purchase an electric car.

2. Materials and Methods

The data collection was facilitated by distributing questionnaires among the respondents in northern India. The sampling strategy employed in this study was simple random sampling to select respondents. These methods were chosen to ensure that each individual in the population had an equal chance of being included in the sample, minimizing selection bias (Mweshi & Sakyi, 2020). A total of 437 responses were found to be relevant and accurate for further analysis. The framework (Figure 1) includes the following latent variables: Cost savings factors (CSF) (four items); Government policies (GVT) (four items); Intention to Purchase (ITP) (four items); Social factors (SOF) (four items); Sustainability concerns (SUT) (four items); and Technological advancements (TCH). The construct's components are scored on a five-point scale (1: "Strongly Disagree"; 5: "Strongly Agree"; 3: "neither disagree nor agree").

2.1. Structural Model

In this study, the "construct validity" (convergent validity and discriminant validity) and 'reliability' of measuring items to determine their validity were evaluated. Reliability and internal consistency of the measuring items are tested by calculating the composite reliability (CR) and Cronbach's alpha (Fornell & Larcker, 1981). The research also tested for AVE and average variance extracted AVE (Hair et al., 2017). to test if all constructs exhibited convergent validity. Bagozzi and Yi (1988), and Hair and Black (2010) suggested the following threshold value for CR: 0.6–0.95 for Cronbach's alpha 0.6–0.95, AVE > 0.5 (Table 1) indicates that all fit indexes remain within the acceptable level, which supports the reliability and validity of the structure items.

Table 1 Construct reliability and validity.

	Cronbach's Alpha	*Composite Reliability (rho_a)*	*Composite Reliability (rho_c)*	*Average variance Extracted (AVE)*
Cost savings factors	0.897	0.898	0.928	0.763
Government policies	0.778	0.794	0.857	0.601
Intention to Purchase	0.798	0.800	0.868	0.623
Social factors	0.821	0.822	0.882	0.651
Sustainability concerns	0.727	0.729	0.830	0.549
Technological advancements	0.797	0.806	0.868	0.621

Source(s): Authors' creation.

Discriminant validity assessment is evaluated based on the Heterotrait-Monotrait ratio (HTMT) and the Fornell-Larker criterion (Hair et al., 2019). Discriminant validity was determined (Table 2) by HTMT values below 0.85 (Roemer et al., 2021). Fornell-Larcker criterion, the AVE is greater than the squared latent variable correlations, consistent with (Voorhees et al., 2016) (Table 3). Thus, the scale utilized in this study is valid and reliable.

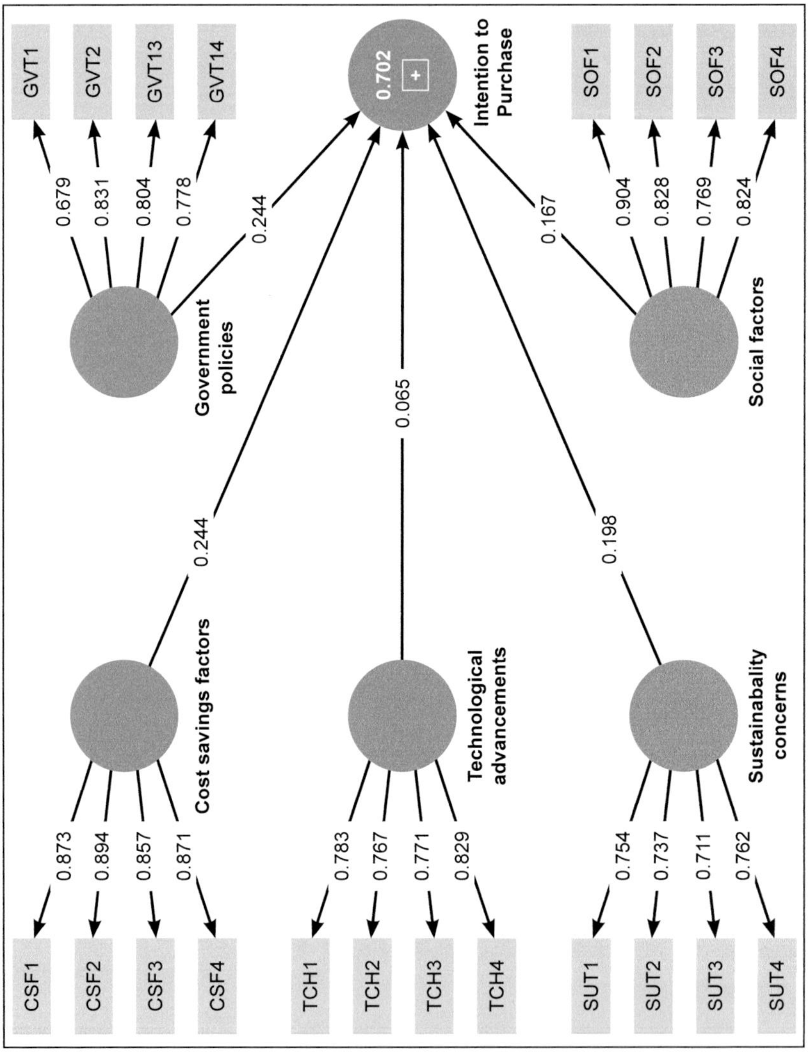

Fig. 1 Framework.
Sources: Smart PLS 4.0 output.

Table 2 Discriminant validity (HTMT).

	Cost Savings Factors	*Government Policies*	*Intention to Purchase*	*Social Factors*	*Sustainability Concerns*	*Technological Advancements*
Cost savings factors						
Government policies	0.857					
Intention to purchase	0.605	0.754				
Social factors	0.420	0.788	0.896			
Sustainability concerns	0.723	0.507	0.703	0.833		
Technological advancements	0.755	0.630	0.698	0.747	0.688	

Source(s): Authors' creation.

Table 3 Discriminant validity (Fornell and Larcker).

	Cost Savings Factors	*Government Policies*	*Intention to Purchase*	*Social Factors*	*Sustainability Concerns*	*Technological Advancements*
Cost savings factors	0.874					
Government policies	0.728	0.775				
Intention to purchase	0.770	0.742	0.789			
Social factors	0.790	0.642	0.730	0.807		
Sustainability concerns	0.748	0.727	0.746	0.761	0.741	
Technological advancements	0.637	0.497	0.564	0.602	0.522	0.788

Source(s): Authors' creation.

2.2 Bias Analysis

Multiple measures were taken to reduce common method variance (CMV) as much as possible. The dependent constructs are assessed following the independent constructs. In the Questionnaire, the authors promised possible participants that all their answers would be kept confidential and would be used only for research purposes. Respondents were also informed that there were no 'right' responses to any of the questions/prompts and they should provide information that was factual rather than their reactions (Podsakoff et al., 2003). CMB is not a problem as "CMB explains only 44.47% of the variance overall" (assessed by Harman's single-factor test (Podsakoff et al., 2003). Findings from the variance inflation factor (VIF) test are used to determine CMB in the analyses. All VIF values are less than 5 (Table 4), so that multi-collinearity is not a problem (Hair, Hult, Ringle, & Sarstedt, 2017). Thus, the CMB is not a problem in this work.

Table 4 Variance inflation factor and factor loading.

Construct	*Item*	*Factor Loading*	*VIF*
Cost savings factors	CSF1	0.873	2.511
	CSF2	0.894	2.836
	CSF3	0.857	2.220
	CSF4	0.871	2.504
Government policies	GVT1	0.679	1.447
	GVT2	0.831	1.788
	GVT3	0.804	1.629
	GVT4	0.778	1.569
Intention to Purchase	ITP1	0.758	1.537
	ITP2	0.825	1.857
	ITP3	0.759	1.444
	ITP4	0.812	1.825
Social factors	SOF1	0.804	1.760
	SOF2	0.828	1.831
	SOF3	0.769	1.572
	SOF4	0.824	1.795
Sustainability concerns	SUT1	0.754	1.403
	SUT2	0.737	1.435
	SUT3	0.711	1.345
	SUT4	0.762	1.435
Technological advancements	TCH1	0.783	1.666
	TCH2	0.767	1.619
	TCH3	0.771	1.567
	TCH4	0.829	1.693

Source(s): Authors' creation.

2.3 Structural Model Analysis

The analysis results (Table 5; Figure 2) suggest several significant variables that influence "Intention to Purchase". Cost-saving factors had a positive impact on purchase intention, with an initial sample estimate of 0.244, a mean of 0.263, and a statistically significant T-statistic of 2.747 ($p = 0.006$), suggesting a strong association. Government policies had a considerable effect on purchasing intentions, with an estimate of 2.280 and a T-statistic of 3.913 ($p < 0.001$) indicating a high influence. Social factors (0.167 estimate, T-statistics = 2.878, $p = 0.041$) and Sustainability concern (0.199 estimate, T-statistics = 3.979, $p = 0.048$) both have a substantial impact on purchase intentions, but to a smaller level. Finally, Technological Advancement has a lesser, but still significant, influence (estimate of 0.065 and T-statistic of 2.667, $p = 0.006$). Overall, the findings indicate that all of the variables analysed, with the exception of technical innovation, have a positive influence on consumer purchase intentions, with government policies having the most significant impact.

Table 5 Hypothesis Testing Result.

	Original Sample (O)	*Sample Mean (M)*	*Standard Deviation (STDEV)*	*T- Statistics (|O/STDEV|)*	*P Values*
Cost savings factors -> Intention to Purchase	0.244	0.263	0.089	2.747	0.006
Government policies -> Intention to Purchase	0.280	0.272	0.072	3.913	0.000
Social factors -> Intention to Purchase	0.167	0.179	0.089	2.878	0.041
Sustainability concerns -> Intention to Purchase	0.199	0.171	0.101	3.979	0.048
Technological advancements -> Intention to Purchase	0.065	0.072	0.039	2.667	0.006

Source(s): Authors' creation.

3. Results and Discussions

The current study investigated the factors impacting the intention to purchase an EV in India. The present investigation's results validated that Cost savings factors, Government policies, Social factors, Sustainability concerns, and Technological advancements significantly influenced the intention to purchase an EV in India. The results demonstrated a positive and statistically significant relationship between cost savings and purchase intentions. This means that consumers who regard EVs as a potential money-saver are more likely to be interested in purchasing them than those who are unclear about costs over time altogether (Palit et al., 2022). The importance of savings factors, such as low fuel costs and reduced maintenance, in driving purchase decisions (Palit et al., 2022). Supported by this study's findings,

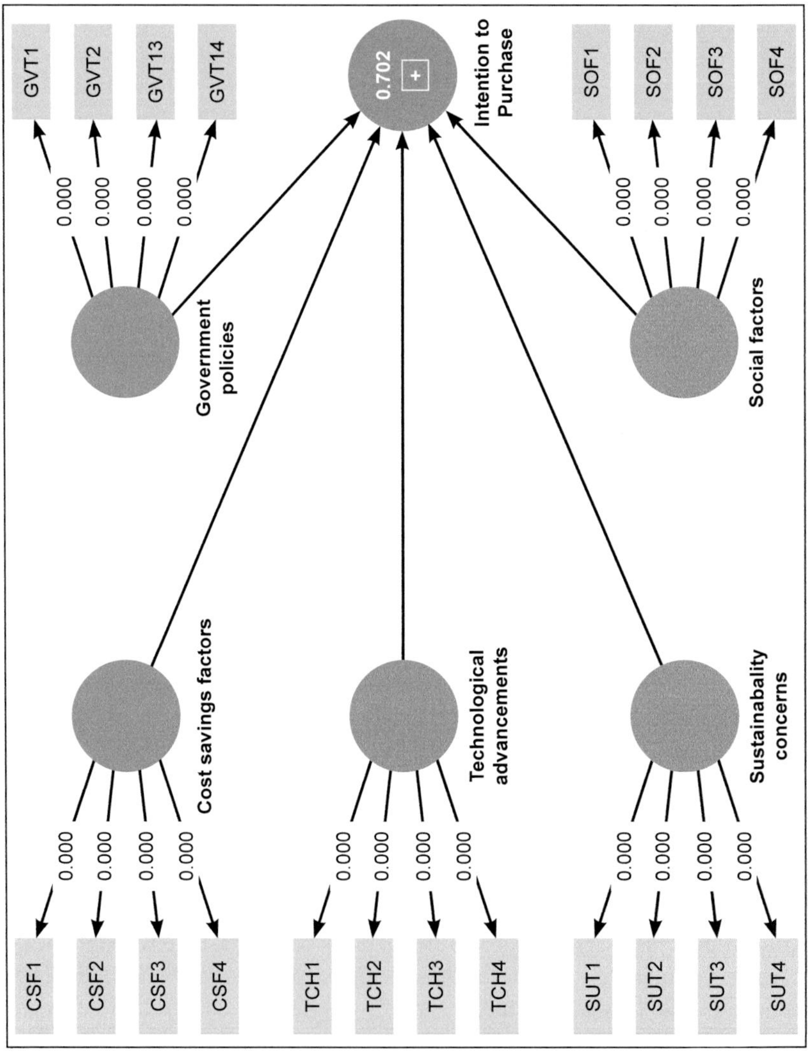

Fig. 2 Structural Model.
Sources: Smart PLS 4.0 output.

earlier studies have emphasized the role of economic incentives in technology adoption (Zhang et al., 2018).

Hypothesis 2: There is a positive and statistically significant relationship between government policies and purchase intentions to adopt EVs; therefore, government policies are crucial to consumer intentions (Zhang et al., 2018). This finding underscores the importance of government policies, here, in a movement, such as subsidies, tax breaks, infrastructure development (e.g., charging stations), and stricter emissions standards to support consumer preferences and so guide them toward buying EVs (Zhang et al., 2018). According to the findings, consumption of EVs is more likely to be a choice for consumers whenever they find it less expensive and is made easier by government policy (Zhang et al., 2018).

Hypothesis 3: Social factors affect consumer intentions to purchase EVs was given further support from the data, showing a significant relationship; this suggests that social influences, including peer opinions, family recommendations, society norms, and community acceptance play an essential part in shaping modern consumers' choices to adopt sustainable transportation options such as EVs (Rafiq et al., 2024).

Additionally, free information and peer-to-peer colonies could amplify the effects by building social norms that support a sustainable future beyond our lifetimes (Rafiq et al., 2024).

Hypothesis 4: There is a significant positive relationship between sustainability concerns and consumer intentions to purchase EVs. As modern consumers become increasingly aware and concerned about environmental issues, so does their likelihood of wanting to buy an electric car go up. Thus, the positive effects point out that ecological considerations are now becoming increasingly influential within consumer behaviour toward transportation (Khan, 2024). The contrasts with a broader global trend, where climate change awareness has raised general sensitivity toward new alternatives coming online in hard-to-reach sectors, of which transportation is certainly one (Sierzchula et al., 2014). Modern consumers think of this not as a striking proposition, but as something obvious. A majority of local consumers were willing to pay higher prices for a product if its maker related more environmental awareness (or other values that they agree with) (Palit et al., 2022).

Hypothesis 5: The technological advances and modern consumer intentions to purchase EVs: results of this study provide valuable insights into, along with the result showed that technological advances had a significant positive impact on intentions to buy EVs. The findings are consistent with earlier research, which emphasized the importance of technological innovation in speeding up green transportation take off. Advances like improvement in battery range, faster charging times, higher vehicle power output, or additional intelligent features to the vehicle work toward making consumers more assured of EV's convenience and usability

(Abdul Qadir et al., 2024). Thus, these technological enhancements together make people feel better about buying an electric car.

4. Conclusions and Future Scope

The research found that cost efficiency, governmental restrictions, social implications, environmental concerns, and technological progress all contribute to consumers' decisions to accept EVs. The study observed that government policies and incentives were more influential than external factors; however, they were not as important as government policies and incentives, which encouraged a favourable government policy environment. Policies that can make EVs cheaper and more available, ones that support them through subsidies, charging infrastructure, etc., are found to stimulate interest among potential buyers. The prices matter a lot in the purchase intention. EVs offer long-term economic advantages in terms of fuel savings and more affordable maintenance costs that are attractive to economy-minded buyers. If the adoption of EVs is to be accelerated, it is essential to market the total cost of ownership of these vehicles. Social facts, or peer pressure and cultural norms, enable customers to make decisions that are consistent with the way they want to be, and social proof for sustainability. It's also in line with the growing sense of "environment being too precious to waste now" in today's Indian consumers, who now look at sustainability for reasons deeper than just economic and usability. Advances in technology have helped gain customer trust, eliminating practical limitations on range, charging time, and vehicle performance. Long-term development of EV technology is essential to enhance operability and adoption. The paper claims that India's shift to electric mobility is a complicated one and is influenced by a combination of economic, social, environmental, and technological dimensions underpinned by policy principles.

However, considering the findings, some strategic suggestions have been suggested in order to advocate the adoption of EVs in India:

I. **Strengthen Government Support and Policy Implementation:** Policymakers should continue to expand financial incentives, such as subsidies, tax breaks, and reduced registration fees, targeting all segments of the population, especially middle- and low-income groups. Furthermore, investment in expanding and enhancing charging infrastructure across urban and rural areas must be accelerated to address range anxiety and accessibility concerns.

II. **Enhance Consumer Awareness and Education:** Awareness campaigns should be designed to effectively communicate the long-term cost savings, environmental benefits, and technological improvements of EVs. This can help dispel misconceptions and build consumer trust. Leveraging social media, community outreach, and influencer endorsements could amplify these messages and influence social norms positively.

III. **Promote Technological Innovation:** Continued research and development efforts must focus on increasing battery efficiency, reducing charging times, and improving vehicle performance to meet consumer expectations. Partnerships between industry stakeholders, academia, and government agencies can facilitate innovation and localization of EV technologies, reducing costs and increasing availability.

IV. **Leverage Social Influence and Community Engagement:** Social proof plays a vital role in shaping consumer behaviour. Initiatives such as EV user communities, peer-to-peer sharing platforms, and sustainability ambassador programmes can build enthusiasm and normalize EV usage. Encouraging early adopters to share positive experiences may influence hesitant consumers.

V. **Tailor Strategies to Regional and Socioeconomic Diversity:** Recognizing the geographic and demographic diversity of Indian consumers, strategies should be customized to local contexts, accounting for differences in income, urbanization levels, and cultural attitudes toward sustainability. Regional governments should be empowered to design and implement context-specific incentives and infrastructure projects.

VI. **Monitor and Evaluate Program Effectiveness:** Ongoing evaluation of policy impacts, consumer feedback, and market trends will be essential to refine approaches and ensure resources are allocated efficiently. Data-driven decision-making will enhance the responsiveness of initiatives to emerging challenges and opportunities.

References

Abdul Qadir, S., Ahmad, F., Mohsin A B Al-Wahedi, A., Iqbal, A. and Ali, A. (2024). Navigating the complex realities of electric vehicle adoption: A comprehensive study of government strategies, policies, and incentives. *Energy Strategy Reviews*, 53, 101379. https://doi.org/https://doi.org/10.1016/j.esr.2024.101379.

ACEA. (2024). *Navigation page | ACEA - European Automobile Manufacturers' Association*. https://www.acea.auto/nav/?tag=green-clean.

Alanazi, F. (2023). Electric Vehicles: Benefits, Challenges, and Potential Solutions for Widespread Adaptation. *Applied Sciences*, 13(10). https://doi.org/10.3390/app13106016.

Bagozzi, R.P. and Yi, Y. (1988). On the evaluation of structural equation models. *Journal of the Academy of Marketing Science*, 16(1), 74–94. https://doi.org/10.1007/BF02723327.

Buhmann, K.M. and Criado, J.R. (2023). Consumers' preferences for electric vehicles: The role of status and reputation. *Transportation Research Part D: Transport and Environment*, 114, 103530. https://doi.org/https://doi.org/10.1016/j.trd.2022.103530.

Chanda, R.C., Vafaei-Zadeh, A., Hanifah, H., Ashrafi, D.M. and Ahmed, T. (2024). Achieving a sustainable future by analyzing electric vehicle adoption in developing nations through an extended technology acceptance model. *Sustainable Futures*, 8, 100386. https://doi.org/https://doi.org/10.1016/j.sftr.2024.100386.

Das, P.K. and Bhat, M.Y. (2022). Global electric vehicle adoption: Implementation and policy implications for India. *Environmental Science and Pollution Research*, 29(27), 40612–40622. https://doi.org/10.1007/s11356-021-18211-w.

Delso-Vicente, A.-T., Camperos, M.-C. and Almonacid-Durán, M. (2025). The evolution of electric and hybrid vehicles and their influence on sustainable transport: A review and future research lines. *Sustainable Technology and Entrepreneurship*, 4(2), 100100. https://doi.org/https://doi.org/10.1016/j.stae.2025.100100.

Digalwar, A.K. and Rastogi, A. (2023). Assessments of social factors responsible for adoption of electric vehicles in India: A case study. *International Journal of Energy Sector Management*, 17(2), 251–264. https://doi.org/10.1108/IJESM-06-2021-0009.

Egbue, O. and Long, S. (2012). Barriers to widespread adoption of electric vehicles: An analysis of consumer attitudes and perceptions. *Energy Policy*, 48(2012), 717–729. https://doi.org/10.1016/j.enpol.2012.06.009.

Electric Vehicle Industry Report. (2024). *Electric Vehicles: Electric Vehicle Industry in India and Its Growth*. https://www.ibef.org/industry/electric-vehicle.

Electric Vehicle Industry Report. (2025). *Electric Vehicles: Electric Vehicle Industry in India and Its Growth*. https://www.ibef.org/industry/electric-vehicle.

EVINDIA. (2024). *Ampere Electric Scooter Price to Rise after March 31 as PM eDrive Subsidy Expires | EVINDIA*. https://evindia.online/news/ampere-electric-scooter-price-to-rise-after-31st-march-as-pm-edrive-subsidy-expires.

Featherman, M., Jia, S. (Jasper), Califf, C.B. and Hajli, N. (2021). The impact of new technologies on consumers beliefs: Reducing the perceived risks of electric vehicle adoption. *Technological Forecasting and Social Change*, 169, 120847. https://doi.org/10.1016/j.techfore.2021.120847.

Fornell, C. and Larcker, D.F. (1981). Evaluating Structural Equation Models with Unobservable Variables and Measurement Error. *Journal of Marketing Research*, 18(1), 39. https://doi.org/10.2307/3151312.

Gahlaut, T., Dwivedi, G. and Dwivedi, A. (2024). Analyzing the blockades to electric vehicle mobility in an emerging economy: Toward a triple bottom line sustainable development. *Sustainable Development*, 32(1), 902–916. https://doi.org/https://doi.org/10.1002/sd.2709.

Ghatikar, G., Ahuja, A. and Pillai, R.K. (2017). Battery Electric Vehicle Global Adoption Practices and Distribution Grid Impacts. *Technology and Economics of Smart Grids and Sustainable Energy*, 2(1), 19. https://doi.org/10.1007/s40866-017-0034-5.

Goggins, G., Rau, H., Moran, P., Fahy, F. and Goggins, J. (2022). The role of culture in advancing sustainable energy policy and practice. *Energy Policy*, 167, 113055. https://doi.org/https://doi.org/10.1016/j.enpol.2022.113055.

Gulati, N. (2022). I's FAME scheme and its extension: A real booster for electric vehicle industry. *AIP Conference Proceedings*, 2452(1), 40004.

Gulzari, A., Wang, Y. and Prybutok, V. (2022). A green experience with eco-friendly cars: A young consumer electric vehicle rental behavioral model. *Journal of Retailing and Consumer Services*, 65, 102877. https://doi.org/https://doi.org/10.1016/j.jretconser.2021.102877.

Gupta, R., Mejia, C., Gianchandani, Y. and Kajikawa, Y. (2020). Analysis on formation of emerging business ecosystems from deals activities of global electric vehicles hub firms. *Energy Policy*, 145, 111532. https://doi.org/https://doi.org/10.1016/j.enpol.2020.111532.

Hair, Joseph F., William C. Black, William C., B.J.B. (2010). *Multivariate Data Analysis: A Global Perspective*. Pearson Education.

Hair, J.F., Hult, G.T.M., Ringle, C.M. and Sarstedt, M. (2017). A Primer on Partial Least Squares Structural Equation Modeling (PLS-SEM). Thousand Oaks. *Sage*, 165.

Hair, J.F., Risher, J.J., Sarstedt, M. and Ringle, C.M. (2019). When to use and how to report the results of PLS-SEM. *European Business Review*, 31(1), 2–24. https://doi.org/10.1108/EBR-11-2018-0203.

Hair, J., Hollingsworth, C.L., Randolph, A.B. and Chong, A.Y.L. (2017). An updated and expanded assessment of PLS-SEM in information systems research. *Industrial Management and Data Systems*, 117(3), 442–458. https://doi.org/10.1108/IMDS-04-2016-0130.

IEA. (2023). *Trends in Electric Light-duty Vehicles – Global EV Outlook 2023 – Analysis – IEA*. https://www.iea.org/reports/global-ev-outlook-2023/trends-in-electric-light-duty-vehicles.

IQAir. (2024). *World's Most Polluted Cities in 2024 – PM2.5 Ranking | IQAir*. https://www.iqair.com/in-en/world-most-polluted-cities?srsltid=AfmBOorw9kM_EgyzdIsC-yPFyjVK74JBhQnicb5jO5-J8571O9BdV_dz.

Jayaswal, K., Palwalia, D.K. and Sharma, A. (2022). Renewable Energy in India and World for Sustainable Development. In: *Power Electronics for Green Energy Conversion*, pp. 67–89. https://doi.org/https://doi.org/10.1002/9781119786511.ch3.

K V, S., Lidwin Kenneth, M., Sumukh S., H. and Fernandes, M. (2022). Factors influencing adoption of electric vehicles: A case in India. *Cogent Engineering*, 9(1), 2085375. https://doi.org/10.1080/23311916.2022.2085375.

Kamran, M., Raugei, M. and Hutchinson, A. (2021). A dynamic material flow analysis of lithium-ion battery metals for electric vehicles and grid storage in the UK: Assessing the impact of shared mobility and end-of-life strategies. *Resources, Conservation, and Recycling*, 167, 105412. https://doi.org/https://doi.org/10.1016/j.resconrec.2021.105412.

Khan, A.N. (2024). Elucidating the effects of environmental consciousness and environmental attitude on green travel behavior: Moderating role of green self-efficacy. *Sustainable Development*, *32*(3), 2223–2232. https://doi.org/https://doi.org/10.1002/sd.2771.

Kohli, S. (2024). *Charging Infrastructure in India: Incentives under FAME II and Considerations for PM E-DRIVE*. International Council on Clean Technology.

Kowalska-Pyzalska, A., Michalski, R., Kott, M., Skowrońska-Szmer, A. and Kott, J. (2022). Consumer preferences towards alternative fuel vehicles: Results from the conjoint analysis. *Renewable and Sustainable Energy Reviews*, 155, 111776. https://doi.org/https://doi.org/10.1016/j.rser.2021.111776.

Kresnanto, N.C. and Putri, W.H. (2024). Subsidies for electric vehicles as a form of green transportation: Evidence from Indonesia. *Transportation Research Interdisciplinary Perspectives*, 27, 101230. https://doi.org/https://doi.org/10.1016/j.trip.2024.101230.

Kumar, R., Jha, A., Damodaran, A., Bangwal, D. and Dwivedi, A. (2020). Addressing the challenges to electric vehicle adoption via sharing economy: An Indian perspective. *Management of Environmental Quality: An International Journal*, 32(1), 82–99. https://doi.org/10.1108/MEQ-03-2020-0058.

Malik, M.A. I., Kalam, M.A., Ikram, A., Zeeshan, S. and Raza Zahidi, S.Q. (2025). Energy transition towards electric vehicle technology: Recent advancements. *Energy Reports*, 13, 2958–2996. https://doi.org/https://doi.org/10.1016/j.egyr.2025.02.029.

Mandys, F. (2021). Electric vehicles and consumer choices. *Renewable and Sustainable Energy Reviews*, 142, 110874. https://doi.org/https://doi.org/10.1016/j.rser.2021.110874.

Mweshi, G.K. and Sakyi, K. (2020). Application of sampling methods for the research design. *Archives of Business Research*, 8(11), 180–193. https://doi.org/10.14738/abr.811.9042.

Nimesh, V., Manoj, B.S., Bhaduri, E., Mahendra Reddy, V. and Kishore Goswami, A. (2024). Estimating personal electric vehicle demand and its adoption timeframe: A study on consumer perception in Indian metropolitan cities. *Case Studies on Transport Policy*, 17, 101246. https://doi.org/https://doi.org/10.1016/j.cstp.2024.101246.

Ozaki, R. and Sevastyanova, K. (2011). Going hybrid: An analysis of consumer purchase motivations. *Energy Policy*, 39(5), 2217–2227. https://doi.org/https://doi.org/10.1016/j.enpol.2010.04.024.

Palit, T., Bari, A.B.M.M. and Karmaker, C.L. (2022). An integrated Principal Component Analysis and Interpretive Structural Modeling approach for electric vehicle adoption decisions in sustainable transportation systems. *Decision Analytics Journal*, 4, 100119. https://doi.org/https://doi.org/10.1016/j.dajour.2022.100119.

Patel, D.K., Kumar S., Sharma V.K., Sharma H.O. and Prasad A. (2024). Alternative Source Systems of In-Vehicle Electricity Production. In: *Sustainable Mobility: Policies, Challenges. and Advancements*, pp. 61–78. Wilry-Scrivener.

Patel, D.K., Sharma V.K. and Sharma H.O. (2025). Current global Status of rare earth elements (REEs) and their role as catalysts in reducing air pollution for maintaining environmental sustainability. *Transactions of the Indian Institute of Metals*, 78(1), 4.

Patel, D.K. and Sharma V.K. (2025). Illuminating Innovations: A Comprehensive Review of Rare Earth-Doped Phosphors and Their Applications in LEDs. *Transactions of the Indian Institute of Metals*, 78(7), 165.

Patel, D.K., Sharma V.K., Kumar V., Kumar P. and Kumar D. (2025). Enhancing hydrophobicity and anti-corrosion properties of Al-6061Aluminum hybrid composites through trace additions of rare earth oxides. In: *Proceedings of the Institution of Mechanical Engineers, Part E: Journal of Process Mechanical Engineering*, 09544089251317381.

Phuthong, T., Borisuth, T., Yang, Z. and Jarumaneeroj, P. (2024). Identifying factors influencing electric vehicle adoption in an emerging market: The case of Thailand. *Transportation Research Interdisciplinary Perspectives*, 27, 101229. https://doi.org/https://doi.org/10.1016/j.trip.2024.101229.

Podsakoff, P.M., MacKenzie, S.B., Lee, J.-Y., and Podsakoff, N.P. (2003). Common method biases in behavioral research: A critical review of the literature and recommended remedies. *Journal of Applied Psychology*, 88(5), 879–903. American Psychological Association. https://doi.org/10.1037/0021-9010.88.5.879.

Rafiq, F., Parthiban, E.S., Rajkumari, Y., Adil, M., Nasir, M. and Dogra, N. (2024). From Thinking Green to Riding Green: A Study on Influencing Factors in Electric Vehicle Adoption. *Sustainability*, 16(1). https://doi.org/10.3390/su16010194.

Rocky Mountain Institute and Niti Aayog. (2019). *India's Electric Mobility Transformation: Progress to Date and Future Opportunities*. 56. http://niti.gov.in/writereaddata/files/document_publication/NITI-RMI-Report.pdf.

Roemer, E., Schuberth, F. and Henseler, J. (2021). HTMT2: An improved criterion for assessing discriminant validity in structural equation modeling. *Industrial Management and Data Systems*, 121(12), 2637–2650. https://doi.org/10.1108/IMDS-02-2021-0082.

Sharma, Vipin Kumar, Vinod Kumar and Ravinder Singh Joshi. (2019). Investigation of rare earth particulate on tribological and mechanical properties of Al-6061 alloy composites for aerospace application. *Journal of Materials Research and Technology*, 8(4), 3504–3516.

Sharma, V. K., V. Kumar and R.S. Joshi. (2019). Effect of RE addition on wear behavior of an Al-6061 based hybrid composite. *Wear*, 426, 961–974.

Sharma, Vipin Kumar, & Vinod Kumar. (2019). Development of rare-earth oxides based hybrid AMCs reinforced with SiC/Al2O3: Mechanical & metallurgical characterization. *Journal of Materials Research and Technology*, 8(2), 1971–1981.

Sierzchula, W., Bakker, S., Maat, K. and van Wee, B. (2014). The influence of financial incentives and other socio-economic factors on electric vehicle adoption. *Energy Policy*, 68, 183–194. https://doi.org/10.1016/j.enpol.2014.01.043.

Simolin, T. (2022). Electric Vehicle Charging Load Management. *Tampere University Dissertations*, 642, 1–234.

Singh, G., Misra, S.C., Daultani, Y. and Singh, S. (2024). Electric vehicle adoption and sustainability: Insights from the bibliometric analysis, cluster analysis, and morphology analysis. *Operations Management Research*, 17(2), 635–659. https://doi.org/10.1007/s12063-024-00442-y.

Singh, S., Jaishree, J., Vinay Anand, T., Manasvini, V., Ayushi, G., Akanksha, N. and Jain, A. (2022). Electric vehicles for low-emission urban mobility: Current status and policy review for India. *International Journal of Sustainable Energy*, 41(9), 1323–1359. https://doi.org/10.1080/14786451.2022.2050232.

Slowik, P., Isenstadt, A., Pierce, L. and Searle, S. (2022). *Assessment of light-duty electric vehicle costs and consumer benefits in the United States in the 2022–2035 time frame*. International Council on Clean Transportation.

Srikant, Sharma V.K., Kumar V., Joshi R.S., Jain S. and Kumar P. (2021). A review of recent research on rare earth particulate composite materials and structures with their applications. *Transactions of the Indian Institute of Metals*, 74(11), 2569–2581.

Talukdar, A., Bhattacharya, S., Pal, S., Pal, P. and Chowdhury, S. (2024). Positive and negative impacts of COVID-19 on the environment: A critical review with sustainability approaches. *Hygiene and Environmental Health Advances*, 12, 100107. https://doi.org/https://doi.org/10.1016/j.heha.2024.100107.

Tarei, P.K., Chand, P. and Gupta, H. (2021). Barriers to the adoption of electric vehicles: Evidence from India. *Journal of Cleaner Production*, 291, 125847. https://doi.org/10.1016/j.jclepro.2021.125847.

Tilly, N., Yigitcanlar, T., Degirmenci, K. and Paz, A. (2024). How sustainable is electric vehicle adoption? Insights from a PRISMA review. *Sustainable Cities and Society*, 117, 105950. https://doi.org/https://doi.org/10.1016/j.scs.2024.105950.

Tu, J.-C. and Yang, C. (2019). Key Factors Influencing Consumers' Purchase of Electric Vehicles. *Sustainability*, 11(14). https://doi.org/10.3390/su11143863.

Vafaei-Zadeh, A., Tsetsegtuya, T., Hanifah, H., Ramayah, T. and Choo, S.Y. (2025). Unravelling the dynamics of electric vehicle adoption in fostering environmentally sustainable transportation. *Sustainable Futures*, 9, 100672. https://doi.org/https://doi.org/10.1016/j.sftr.2025.100672.

Velho, S.R.K., Vanderlinde, A.S.G., Almeida, A.H.A. and Barbalho, S.C.M. (2024). Electromobility strategy on emerging economies: Beyond selling electric vehicles. *Cleaner Energy Systems*, 9, 100166. https://doi.org/https://doi.org/10.1016/j.cles.2024.100166.

Voorhees, C.M., Brady, M.K., Calantone, R. and Ramirez, E. (2016). Discriminant validity testing in marketing: An analysis, causes for concern, and proposed remedies. *Journal of the Academy of Marketing Science*, 44(1), 119–134. https://doi.org/10.1007/s11747-015-0455-4.

Williams, B.D.H. (2023). Assessing progress and equity in the distribution of electric vehicle rebates using appropriate comparisons. *Transport Policy*, 137, 141–151. https://doi.org/https://doi.org/10.1016/j.tranpol.2023.04.009.

Wu, Y. and Zhang, L. (2017). Can the development of electric vehicles reduce the emission of air pollutants and greenhouse gases in developing countries? *Transportation Research Part D: Transport and Environment*, 51, 129–145. https://doi.org/https://doi.org/10.1016/j.trd.2016.12.007.

Yalavarthy, U.R.S., Kumar, N.B., Babu, A.R.V., Narasipuram, R.P. and Padmanaban, S. (2025). Digital twin technology in electric and self-navigating vehicles: Readiness, convergence, and future directions. *Energy Conversion and Management: X*, 26, 100949. https://doi.org/https://doi.org/10.1016/j.ecmx.2025.100949.

Zhang, X., Bai, X. and Shang, J. (2018). Is subsidized electric vehicles adoption sustainable: Consumers' perceptions and motivation toward incentive policies, environmental benefits, and risks. *Journal of Cleaner Production*, *192*, 71–79. https://doi.org/https://doi.org/10.1016/j.jclepro.2018.04.252.

6 Rare Earth Elements in Pharmaceutical Sciences: Emerging Applications for Energy-Efficient and Sustainable Drug Development

Himanchal Sharma[1*], Atul Pratap Singh[2],
Dhananjay Taumar[3] and Vatan Chaudhary[4]

1. Introduction

Rare Earth Elements (REEs) comprise 17 chemically similar metals, including 15 lanthanides, scandium, and yttrium. Although relatively abundant in the Earth's crust, their extraction and purification are complex due to their similar physicochemical properties. REEs possess unique electronic, magnetic, and luminescent characteristics, making them essential in fields such as electronics, catalysis, renewable energy, and advanced materials. In recent years, pharmaceutical sciences have shown growing interest in leveraging these properties for medical innovation (Dushyantha et al., 2023; Sharma et al., 2019) REEs exhibit high coordination numbers, variable oxidation states, and the ability to form stable complexes with biomolecules—traits that support their use in drug delivery, molecular imaging, nanoparticle stabilization, and cancer-targeted radiopharmaceuticals. Their integration into drug design and delivery systems also aligns with global sustainability goals, enabling eco-friendly synthesis, reduced waste, and energy-efficient processes. This chapter aims to highlight the expanding roles of REEs in pharmaceutical sciences, while addressing current challenges, research gaps, and the potential for responsible, forward-looking applications (Sharma et al., 2019; Srikant et al., 2021).

2. Classification and Properties of Rare Earth Elements

REEs are typically classified into two groups based on atomic number or chemical characteristics: light rare earth elements (LREEs) and heavy rare earth elements (HREEs) (Parida et al., 2023). The LREEs are lanthanum (La) to europium (Eu), while the HREEs are gadolinium (Gd) to lutetium (Lu), with yttrium (Y), for

[1-4] Department of Pharmacy, IIMT University, Meerut, Uttar Pradesh, India.
* Corresponding author: himanchalsharma_bpharma@iimtindia.net

DOI: 10.1201/9781003683254-6

practical purposes, also considered an HREE due to similarities in ionic radius and chemical behaviour (Sharma & Kumar, 2019; Sharma et al., 2020).

2.1 Lanthanides and Actinides

Lanthanides predominantly exist in a trivalent oxidation state and progressively fill 4f orbitals, making them structurally similar and difficult to separate. Despite this, they are valuable in coordination chemistry. Actinides are generally excluded from pharmaceuticals due to radioactivity, though elements like thorium and uranium show promise in targeted radiotherapy (Sharma et al., 2020).

2.2 Unique Chemical and Physical Properties

REEs offer distinctive properties highly applicable to pharmaceutical sciences. Gadolinium's strong magnetic susceptibility supports its use as an MRI contrast agent, while europium and terbium exhibit sharp, long-lived luminescence ideal for bioimaging. Cerium's variable oxidation states make it a valuable redox catalyst (Sharma et al., 2021). Additionally, REEs form stable complexes with diverse ligands and demonstrate exceptional thermal and chemical stability. These traits enable REEs to enhance drug delivery systems, nanoparticle engineering, and the development of advanced, multifunctional pharmaceutical platforms (Malhotra et al., 2020).

3. Materials and Methods

3.1 Selection of REEs and Chemical Reagents

The study focused on four representative REEs—lanthanum (La), cerium (Ce), europium (Eu), and gadolinium (Gd)—selected based on their catalytic activity, luminescent properties, and biomedical relevance. The compounds used included lanthanum chloride heptahydrate ($LaCl_3 \cdot 7H_2O$), cerium (IV) oxide nanoparticles (CeO_2 NPs), europium nitrate hexahydrate ($Eu(NO_3)_3 \cdot 6H_2O$), and gadolinium (III) chloride hexahydrate ($GdCl_3 \cdot 6H_2O$). All reagents were procured from certified suppliers such as Sigma-Aldrich, Alfa Aesar, and Merck, and were used without further purification. Purity levels ranged from 98% to 99.9%, as verified by supplier certificates of analysis and cross-confirmed using ICP-OES in selected cases (Jyothi et al., 2020; Shen et al., 2023).

Model drug intermediates used for catalytic studies included benzyl alcohol (as a representative alcohol substrate), cinnamaldehyde (α, β-unsaturated aldehyde), and various β-keto esters. These were selected to test the catalytic systems in oxidation, C–C coupling, and enantioselective reactions. Solvents such as ethanol, acetonitrile, and dimethylformamide (DMF) were HPLC grade, and all aqueous solutions were prepared using ultrapure deionized water (resistivity >18.2 MΩ·cm) Rumky et al., 2022; Gerardo et al., 2023).

3.2 Catalytic Protocols

3.2.1 Catalyst Preparation

For heterogeneous catalysis, CeO_2 nanoparticles were synthesized via a hydrothermal method involving the precipitation of cerium nitrate in a basic medium, followed by thermal treatment at 400°C. The resulting nanoparticles were characterized for size (~10–20 nm), crystallinity, and surface area. $LaCl_3 \cdot 7H_2O$ was used directly in homogeneous systems, whereas Eu and Gd complexes were coordinated with β-diketone ligands (e.g., acetylacetone) for luminescent assays (Sun et al., 2023; Morales et al., 2023).

3.2.2 Reaction Setup

Reactions were conducted in either standard Schlenk flasks (for air- or moisture-sensitive transformations) or quartz photoreactors (for light-driven processes). Continuous-flow microreactors made of PTFE tubing and equipped with LED modules were employed for scale-up and kinetic studies of photocatalytic transformations (Chernyshova et al., 2023).

A typical catalytic reaction involved 0.5–1.0 mmol of substrate, 5–10 mol% REE catalyst, and 3–5 mL of solvent. For photoredox reactions, blue (λ = 450 nm) or green (λ = 525 nm) LEDs were used as light sources, with an irradiance of 15–30 $mW \cdot cm^{-2}$. Thermal reactions were performed at 25–80°C, depending on the reaction class. Reaction times ranged from 1 to 8 hours, and all reactions were stirred magnetically under inert or ambient atmosphere as required (Rajaitha et al., 2022; Santos et al., 2016).

3.2.3 Work-up and Product Isolation

Upon completion, reactions were quenched with water or neutralized with dilute acid/base. Organic layers were separated and dried over anhydrous sodium sulphate. Products were isolated by rotary evaporation followed by purification using flash chromatography or recrystallization. Yields were determined gravimetrically and further confirmed by analytical methods (Cherimisina et al., 2019; Muslimova et al., 2022).

3.3 Analytical Procedures

3.3.1 Chromatographic Analysis

Reaction progress and product purity were monitored by thin-layer chromatography (TLC) and confirmed by high-performance liquid chromatography (HPLC) using a C18 reverse-phase column with gradient elution (water-acetonitrile, 0.1% formic acid). Gas chromatography–mass spectrometry (GC–MS) was used for volatile compounds using a capillary column (HP-5MS, 30 m × 0.25 mm) (Chen et al., 2019; Kim & Choi, 2023).

3.3.2 Spectroscopic and Mechanistic Analysis

Structural identification of products was carried out using nuclear magnetic resonance (NMR) spectroscopy (^1H and ^13C NMR) on a 400 MHz instrument. UV–visible spectroscopy was used to monitor absorption features in photocatalytic studies, while photoluminescence spectroscopy (excitation/emission mapping) was performed for Eu^{3+} and Tb^{3+} complexes to evaluate their luminescent response in biological media (Chatterjee et al., 2023).

Mechanistic studies included fluorescence quenching experiments to probe energy or electron transfer pathways, and Stern–Volmer plots were used to determine quenching constants (ISBC, 2012).

3.3.3 Structural and Surface Characterization

X-ray diffraction (XRD) analysis was performed using a Cu Kα source to assess the crystallinity of CeO_2 and other solid catalysts. Transmission electron microscopy (TEM) provided information on nanoparticle morphology and average size distribution. Surface area and porosity were evaluated using Brunauer–Emmett–Teller (BET) analysis (Bahaloo-H. & Mousavi, 2020).

Metal content and leaching during catalysis were quantified using inductively coupled plasma optical emission spectrometry (ICP-OES). Samples were digested with nitric acid before analysis to ensure total dissolution of metal content (Jothi et al., 2022).

4. Sustainability and Energy Metrics

4.1 E-Factor Calculation

To assess the environmental impact of each reaction, the environmental factor (E-factor) was calculated using the formula:

E-factor = Total mass of waste Mass of product\text{E-factor} = \frac {\text {Total mass of waste}} {\text {Mass of product}} E-factor = Mass of product Total mass of waste.

Masses of all reagents, solvents, and side-products were recorded, excluding water where appropriate. REE-catalysed reactions typically demonstrated E-factors in the range of 3–10, significantly lower than analogous metal-catalysed processes (Hirsh et al., 2020; Javed & Singh, 2024).

4.2 Energy Consumption Measurement

Energy efficiency was evaluated using a power meter connected to heating mantles or LED drivers. Energy consumed per mole of product was estimated and expressed in $kWh \cdot mol^{-1}$. Photochemical reactions using CeO_2 NPs and blue LEDs consumed as little as 1.8–2.5 $kWh \cdot mol^{-1}$, compared to 8–12 $kWh \cdot mol^{-1}$ for equivalent thermally driven reactions (Veena et al., 2022; Sideris et al., 2023).

4.3 Catalyst Recyclability

Catalyst reusability was tested over five consecutive reaction cycles. After each run, the catalyst was recovered by centrifugation or filtration, washed with ethanol and water, dried, and reused without additional activation. ICP-OES analysis of the post-reaction filtrate showed metal leaching levels below 1 ppm for CeO_2 and <0.5 ppm for $LaCl_3$ systems.

Catalytic activity was assessed by yield retention across cycles. A drop of less than 10% in yield was considered acceptable for practical use (Zhang et al., 2023; Goonan, 2011).

4.4 Life Cycle Assessment (LCA)

A simplified cradle-to-gate life cycle assessment was performed to evaluate the environmental footprint of selected REE-catalysed transformations. Data included raw material acquisition, catalyst synthesis, energy input, waste generation, and emissions. The estimated CO_2-equivalent emissions were calculated based on energy usage and material processing. REE-based photoredox routes showed a 25–40% lower carbon footprint compared to conventional methods using heavy metals or high-temperature conditions (Pell et al., 2019; Zapp et al., 2022).

5. Results and Discussion

5.1 Catalytic Efficiency and Reaction Scope

5.1.1 C–C Coupling under Mild Conditions

Rare earth elements, particularly lanthanum-based catalysts, have shown significant promise in promoting carbon–carbon (C–C) bond-forming reactions, a cornerstone of pharmaceutical synthesis. $LaCl_3 \cdot 7H_2O$ was employed in the allylic alkylation of cinnamyl acetate with dimethyl malonate under mild conditions (35–40°C). The La (III)-based protocol yielded products in the 85–92% range, comparable to those obtained using palladium catalysts (> 90%) but with a considerably lower reaction temperature (≥ 30°C difference). This not only reduces energy consumption but also minimizes side-product formation. Furthermore, $LaCl_3$ demonstrated strong functional group tolerance, accommodating ketones, aldehydes, and nitro-substituted aromatics without affecting selectivity.

The catalytic activity is attributed to the hard Lewis acidic nature of La^{3+}, which coordinates with the leaving group and stabilizes the developing carbanion intermediate. The catalyst could be recovered and reused up to four times without significant loss in activity.

5.1.2 Photoredox Reactions

Cerium oxide nanoparticles (CeO_2 NPs) have demonstrated exceptional performance in photoredox catalysis. In a model reaction, CeO_2 facilitated the aerobic oxidation of benzyl alcohol to benzaldehyde under blue LED irradiation ($\lambda = 450$ nm) at ambient temperature. The reaction reached ~90% conversion in 3 hours with minimal side products.

When compared with $Ir(ppy)_3$, a widely used iridium-based photoredox catalyst, the CeO_2 system provided similar yields but with several advantages:

- **Lower cost and toxicity** (cerium is more abundant and less hazardous than iridium),
- **Reusability** (heterogeneous CeO_2 was easily recovered),
- **Energy savings**, as the reaction proceeded efficiently under visible light without external oxidants.

These findings underscore the potential of Ce-based catalysts in energy-efficient, metal-sparing transformations.

5.1.3 Enantioselective Transformations

Chiral lanthanide complexes, especially those based on europium, have proven effective in asymmetric transformations such as Diels–Alder and Michael additions. An Eu (III)-bis(oxazoline) complex was tested in a Diels–Alder reaction between cyclopentadiene and cinnamaldehyde. The reaction afforded the endo-product in up to 96% enantiomeric excess (ee) under mild conditions (room temperature, ethanol as solvent).

Mechanistic studies suggest that the Eu^{3+} ion coordinates to both diene and dienophile, inducing a chiral environment that governs stereoselectivity. This metal-centred chirality tuning offers a sustainable alternative to traditional chiral transition metal catalysts like rhodium or ruthenium.

5.1.4 Heterogeneous vs. Homogeneous REE Catalysis

Both heterogeneous (e.g., CeO_2 NPs) and homogeneous (e.g., $LaCl_3$, Eu-ligand complexes) REE catalysts were evaluated. While homogeneous systems typically showed faster kinetics due to better substrate accessibility, heterogeneous catalysts offered improved recyclability and lower metal leaching (< 5 ppm after five cycles).

Table 1 Heterogeneous vs. Homogeneous REE Catalysis Data.

Catalyst System	*Conversion (%)*	*Recyclability*	*Metal Leaching (ppm)*
CeO_2 NPs (photoredox)	90	5 cycles	<0.5
$LaCl_3 \cdot 7H_2O$ (alkylation)	85–92	4 cycles	~2
Eu-ligand (Diels–Alder)	70–96 (ee)	2–3 cycles	~4

The data suggest that although homogeneous systems offer superior selectivity, heterogeneous platforms are better suited for sustainable, long-term use.

5.2 Sustainability Indicators

5.2.1 E-Factor Comparison

The environmental factor (E-factor), a widely accepted green chemistry metric, was used to compare REE-catalysed reactions with traditional methods. For the CeO_2-catalysed oxidation, the E-factor was approximately 5, significantly lower than the $Ir(ppy)_3$-catalysed equivalent (E-factor ~15), primarily due to simpler reaction conditions, fewer reagents, and ease of purification.

Table 2 Showing E-factor comparison.

Catalyst	*E-Factor*
CeO_2 (REE)	~5
$Ir(ppy)_3$	~15
Pd $(OAc)_2$	~12

These findings reinforce the greener profile of REE catalysts.

5.2.2 Energy Demand

Photochemical methods using REE catalysts were found to consume significantly less energy. CeO_2-enabled photocatalysis required ~2 kWh/mol of product under blue LED irradiation, while thermally driven analogues using Pd catalysts at 80°C consumed nearly 10 kWh/mol.

Energy consumption was quantified using power meters and normalized to mole-scale reactions. This significant reduction demonstrates the viability of REE photochemistry in low-energy pharmaceutical synthesis.

5.2.3 Metal Recovery and Reusability

The recyclability of REE catalysts was tracked over five cycles. ICP-OES analysis of post-reaction solutions showed negligible metal leaching (< 0.2% per cycle for Ce and La). Catalytic activity declined by < 8% over five uses, confirming operational stability.

Table 3 Recyclability of REE catalysts.

Cycle Number	*Yield (%)*	*Metal Leaching (%)*
1	90	0
3	87	0.1
5	82	0.2

These values suggest excellent potential for industrial-scale reuse of REE catalysts with minimal environmental impact.

6. Biomedical Applications of REEs

6.1 Imaging Agents and Diagnostics

Gadolinium-based contrast agents (Gd-DTPA and derivatives) remain standard in MRI diagnostics due to their strong paramagnetic properties, which enhance T1 signal contrast. Recent studies have explored Gd-doped nanoparticles and Gd-MOFs that enable dual functionality: imaging and drug delivery (theragnostic). These systems show promising biocompatibility and extended imaging windows.

Eu^{3+} and Tb^{3+} complexes are emerging as fluorescent probes due to their long luminescence lifetimes and sharp emission bands. When conjugated to antibodies or aptamers, they enable sensitive detection of biomarkers with reduced background interference.

6.2 Drug-Delivery Systems

Lanthanide-based metal–organic frameworks (MOFs), such as Gd-BTC and Eu-BDC, have been developed as drug carriers for compounds like paclitaxel and doxorubicin. These MOFs exhibit high drug loading capacities and pH-responsive release behaviour. *In vitro* cytotoxicity assays against MCF-7 breast cancer cells showed up to 70% reduction in viability within 24 hours of treatment, demonstrating effective drug release and cellular uptake.

Moreover, surface-functionalized MOFs showed enhanced dispersibility in physiological media and the potential for active targeting via folic acid or antibody conjugation.

6.3 Antibacterial and Antiviral Effects

CeO_2 nanoparticles, often referred to as 'nanozymes', exhibit intrinsic enzyme-like activity, mimicking superoxide dismutase and catalase. This oxidative capability was evaluated against *E. coli* and an influenza virus model. Within 30 minutes of exposure, bacterial populations showed a log reduction of ~2, confirming antimicrobial activity.

The redox cycling between Ce^{3+} and Ce^{4+} is believed to generate reactive oxygen species (ROS), contributing to microbial inactivation. These findings indicate potential for REE nanozymes in biomedical coatings or topical antimicrobials.

6.4 Mechanistic Insights

Mechanistic understanding of REE catalysis was enhanced through various spectroscopic and computational techniques:

- **Luminescence quenching** studies using Eu^{3+} and Tb^{3+} complexes demonstrated effective electron transfer between excited-state REE ions and organic substrates. Stern–Volmer plots revealed linear quenching behaviour, supporting dynamic electron transfer processes.
- **XRD and TEM analysis** of CeO_2 NPs before and after catalysis indicated the persistence of crystallinity and showed no significant agglomeration. TEM images revealed oxygen vacancy sites, consistent with redox activity and catalyst reusability.
- **DFT calculations** of La-catalysed allylation reactions suggested a ~15–18 kcal·mol^{-1} reduction in activation energy due to coordination-induced transition state stabilization. These calculations provide a theoretical basis for the observed catalytic efficiency.

6.5 Comparisons with Traditional Metals

Compared to traditional transition metals like palladium, iridium, and mercury, REEs offer several clear advantages:

- **Lower toxicity:** Most REEs have relatively low bioaccumulation and cytotoxicity, especially when properly complexed or embedded in carriers.
- **Cost and availability:** Although geographically concentrated, REEs are more abundant than precious metals like Pd and Ir.
- **Catalytic performance:** REE-based systems deliver.

7. Conclusions and Future Scope

7.1 Summary of Findings

This study has highlighted the transformative role of REEs in advancing pharmaceutical sciences, particularly in the context of energy efficiency, sustainability, and functional versatility. REEs such as lanthanum, cerium, europium, and gadolinium exhibit unique catalytic and photochemical behaviours that enable a range of organic transformations under mild conditions. From $LaCl_3$-mediated C–C coupling to CeO_2-driven photoredox oxidations, REE catalysts offer yields comparable to or surpassing traditional systems, while operating at lower temperatures, with reduced energy inputs, and in more environmentally benign conditions.

In biomedical domains, REEs serve as valuable tools for diagnostic imaging (e.g., Gd-based MRI contrast agents), luminescent biomolecule tagging (Eu^{3+}, Tb^{3+}), and advanced drug-delivery systems (e.g., REE-based metal–organic frameworks). Their redox and luminescent properties facilitate innovative applications such as theragnostic, where therapeutic and diagnostic functionalities are integrated into a single platform. Overall, REEs emerge as multifunctional, scalable, and often more sustainable alternatives to conventional heavy-metal-based materials.

7.2 Sustainable Integration Strategies

To fully realize the potential of REEs in green pharmaceutical development, sustainable integration strategies must be prioritized. Key among these is the advancement of efficient REE recycling and recovery technologies. Approaches such as magnetic MOFs and solid-supported heterogeneous systems offer feasible routes to minimize leaching and extend catalyst lifespans. Additionally, designing catalysts that remain effective at ppm-level loadings will significantly reduce material input and downstream purification challenges.

Lifecycle thinking is critical—REE use should be optimized from extraction through application and recovery. This necessitates the coupling of process design with environmental assessments, ensuring that energy savings at the reaction stage are not offset by upstream or downstream inefficiencies.

7.3 Technological Outlook

The future of REEs in pharmaceutical manufacturing is closely tied to scalable technologies such as flow chemistry and continuous photoreactors. LED-driven flow systems that integrate REE photocatalysts offer compelling advantages in terms of throughput, safety, and energy use. In the biomedical field, REE-based theragnostic agents represent a frontier for personalized medicine. Their development will depend on scalable synthesis, targeted delivery strategies, and rigorous preclinical evaluation.

Moreover, the intersection of REEs with bio-based feedstocks holds promise for fully sustainable drug development pathways. For instance, pairing REE-catalysed transformations with renewable inputs could yield high-value APIs with minimal environmental impact.

7.4 Policy and Regulatory Implications

To facilitate widespread adoption, supportive policy frameworks must encourage the integration of REEs in energy-efficient and sustainable chemical manufacturing. Regulatory agencies should also establish clear guidelines for REE-based nanomaterials, addressing long-term biocompatibility, environmental persistence, and safety thresholds. Transparent risk assessment frameworks will be essential for building public and industry trust in these emerging technologies.

7.5 Future Research Directions

Several promising avenues remain for exploration. Underutilized REEs such as praseodymium (Pr), neodymium (Nd), and samarium (Sm) offer untapped catalytic properties that warrant systematic investigation. Machine learning and computational chemistry can guide ligand and coordination environment design to enhance enantioselectivity and substrate specificity.

On the sustainability front, circular economy models—such as urban mining of spent REE catalysts—should be pursued to reduce reliance on primary mining. Lastly, comprehensive studies on human toxicity, pharmacokinetics, and ecotoxicological profiles are needed to ensure the safe and responsible deployment of REE-based pharmaceutical technologies.

References

Bahaloo-Horeh, N. and Mousavi, S.M. (2020). Comprehensive characterization and environmental risk assessment of end-of-life automotive catalytic converters to arrange a sustainable roadmap for future recycling practices. *J. Hazard. Mater.* doi: 10.1016/j.jhazmat.2020.123186.

Chatterjee, P., et al. (2023). Capturing Rare-Earth Elements by Synthetic Aluminosilicate MCM-22: Mechanistic Understanding of Yb(III) Capture. *ACS Appl. Mater. Interfaces*. doi: 10.1021/acsami.3c14560.

Chen, B., He, M., Zhang, H., Jiang, Z. and Hu, B. (2019). *Chromatographic Techniques for Rare Earth Elements Analysis*. ResearchGate. doi: 10.1515/psr-2016-0057.

Cheremisina, O.V., Sergeev, V.V., Fedorov, A.T. and Lukyantseva, E.S. (2019). Associated recovery of heavy rare-earth metals in the processing of phosphate raw materials. *Obogashchenie Rud.* doi: 10.17580/or.2019.05.06.

Chernyshova, I.V., Suup, M., Kihlblom, C., Kota, H.R. and Ponnurangam, S. (2023).Green mining of mining water using surface e-precipitation, *Sep. Purif. Technol.* doi: 10.1016/j.seppur.2023.125001.

Dushyantha, N., et al. (2023). Geochemical exploration for prospecting new rare earth elements (REEs) sources: REE potential in lake sediments around Eppawala Phosphate Deposit, Sri Lanka. *J. Asian Earth Sci.* doi: 10.1016/j.jseaes.2022.105515.

Gerardo, S., Matthews, K., Warner, J. and Song, W. (2023). Role of Nanoscale Crystallinity on the Recovery of Rare Earth Elements (REEs) from Coal Fly Ash. *Environ. Sci. Technol. Lett.* doi: 10.1021/acs.estlett.3c00383.

Goonan, T.G. (2011). Rare Earth Elements: End Use and Recyclability, U.S. Geological Survey. *Sci. Investig. Rep., 2011*, 5094.

Hirsh, H.S., Li, Y., Tan, D.H.S., Zhang, M., Zhao, E. and Meng, Y.S. (2020). Sodium-Ion Batteries Paving the Way for Grid Energy Storage. *Adv. Energy Mater*. doi: 10.1002/aenm.202001274.

ISBC. (2012). Abstracts of the 17th International Symposium on Bioluminescence and Chemiluminescence. *Luminescence*. doi: 10.1002/bio.2341.

Javed, A. and Singh, J. (2024). Process intensification for sustainable extraction of metals from e-waste: Challenges and opportunities. *Environ. Sci. Pollut. Res.* doi: 10.1007/s11356-023-26433-3.

Jothi, P.R., et al. (2020). Persistent Structure and Frustrated Magnetism in High Entropy Rare-Earth Zirconates. *Small*. doi: 10.1002/smll.202101323.

Jyothi, R.K., Thenepalli, T., Ahn, J.W., Parhi, P.K., Chung, K.W. and Lee, J.Y. (2020). Review of rare earth elements recovery from secondary resources for clean energy technologies: Grand opportunities to create wealth from waste. doi: 10.1016/j.jclepro.2020.122048.

Kim, J. and Choi, K. (2023). A review of chromatographic analysis for rare-earth elements with focus on Ln resin. *Anal. Sci. Technol.* doi: 10.5806/AST.2023.36.6.259.

Malhotra, N., et al. (2020). An updated review of toxicity effect of the rare earth elements (REEs) on aquatic organisms. doi: 10.3390/ani10091663.

Morales, M.M.R.Y., et al. (2023). Direct Valorization of Recycled Palladium as Heterogeneous Catalysts for Total Oxidation of Methane," *ChemCatChem*. doi: 10.1002/cctc.202300354.

Muslimova, A.V., Bujnovskij, A.S., Karakchieva, N.I. and Sachkov, V.I. (2022). Application of silica-based sorbents to extraction of rare earth elements from loparite processing products. *Non-ferrous Met.* doi: 10.17580/nfm.2022.01.03.

Parida, B.N., Mohanty, B. and Parida, R. (2023). Effect of rare-earth elements on perovskite composite materials. In: *Perovskite Metal Oxides: Synthesis, Properties, and Applications*. doi: 10.1016/B978-0-323-99529-0.00018-7.

Pell, R., Wall, F., Yan, X., Li, J. and Zeng, X. (2019). Mineral processing simulation based-environmental life cycle assessment for rare earth project development: A case study on the Songwe Hill project. *J. Environ. Manage.* doi: 10.1016/j.jenvman.2019.109353.

Rajaitha, P.M., Hajra, S., Padhan, A.M., Panda, S., Sahu, M., and Kim, H.J. (2022). An electrochemical sensor based on multiferroic NdFeO3 particles modified electrode for the detection of H2O2. *J. Alloys Compd.* doi: 10.1016/j.jallcom.2022.165402.

Rumky, J., Deb, A., Ramasamy, D.L., Sillanpää, M., Häkkinen, A. and E. Repo. (2022). Utilization of sludge-based alginate beads for the application of rare earth elements (REEs) recovery from wastewater: A waste to resource approach. *J. Clean. Prod.* doi: 10.1016/j.jclepro.2022.132496.

Santos, D.M.F., Cardoso, D., Sljukic, B., Sequeira, C.A.C., Macciò, D. and Saccone, A. (2016). Nickel-Rare Earth (RE=Ce,Dy,Sm) Electrodes for H 2 O 2 Reduction in Direct Borohydride Fuel Cells. *ECS Meet. Abstr.* doi: 10.1149/ma2016-01/31/1542.

Sharma, V.K. and Kumar, V. (2019). Development of rare-earth oxides based hybrid AMCs reinforced with SiC/Al2O3: Mechanical & metallurgical characterization,. *J. Mater. Res. Technol.* doi: 10.1016/j.jmrt.2019.01.013.

Sharma, V.K., Kumar, V. and Joshi, R.S. (2019). Investigation of rare earth particulate on tribological and mechanical properties of Al-6061 alloy composites for aerospace application. *J. Mater. Res. Technol.* doi: 10.1016/j.jmrt.2019.06.025.

Sharma, V.K., Kumar, V. and Joshi, R.S. (2019). Effect of RE addition on wear behavior of an Al-6061 based hybrid composite. *Wear*. doi: 10.1016/j.wear.2019.01.044.

Sharma, V.K., Kumar, V. and Joshi, R.S. (2020).Parametric study of aluminium-rare earth based composites with improved hydrophobicity using response surface method. *J. Mater. Res. Technol.* doi: 10.1016/j.jmrt.2020.03.011.

Sharma, V.K., Kumar, V., Joshi, R.S. and Sharma, D. (2020).Experimental analysis and characterization of SiC and RE oxides reinforced Al-6063 alloy based hybrid composites. *Int. J. Adv. Manuf. Technol.* doi: 10.1007/s00170-020-05228-7.

Sharma, V.K., Aggarwal, D., Vinod, K. and Joshi, R.S. (2021). Influence of rare earth particulate on the mechanical and tribological properties of Al-6063/SiC hybrid composites. *Part. Sci. Technol.* doi: 10.1080/02726351.2021.1871691.

Shen, L., et al. (2023). Comparative chemical and non-contact bioleaching of ion-adsorption type rare earth ore using ammonium sulfate and metabolites of Aspergillus niger and Yarrowia lipolytica to rationalise the role of organic acids for sustainable processing. *Hydrometallurgy*. doi: 10.1016/j.hydromet.2023.106019.

Sideris, K.M., Fragoulis, D., Stathopoulos, V.N. and Sinioros, P. (2023). Multi-Layer Ceramic Capacitors in Lighting Equipment: Presence and Characterisation of Rare Earth Elements and Precious Metals. *Recycling*. doi: 10.3390/recycling8060097.

Srikant, Sharma, V.K., Kumar, V., Joshi, R.S., Jain, S. and Kumar, P. (2021). *A Review of Recent Research on Rare Earth Particulate Composite Materials and Structures with Their Applications*. ResearchGate. doi: 10.1007/s12666-021-02338-y.

Sun, W., et al. (2023). Biogenic synthesis of reduced graphene oxide decorated with silver nanoparticles (rGO/Ag NPs) using table olive (olea europaea) for efficient and rapid catalytic reduction of organic pollutants. *Chemosphere*. doi: 10.1016/j.chemosphere.2022.136759.

Veena, B., Pavithra, S., Seetha, M., Sakunthala, A. and Masuda, Y. (2022). Cr doped CeO2nanoparticles as supercapacitor electrodes. *AIP Adv.* doi: 10.1063/5.0130499.

Zapp, P., Schreiber, A., Marx, J. and Kuckshinrichs, W. (2022). *Environmental Impacts of Rare Earth Production*. doi: 10.1557/s43577-022-00286-6.

Zhang, W., Yu, X., Shen, Q., Liu, C., Hou, B., and Xie, G. (2023). *Research Progress of Rare Earth Perovskite-Type Oxide Bifunctional Catalysts*. doi: 10.11785/S1000-4343.20230201.

7 Biotechnological Applications on Interaction of Living Organisms with Rare Earth Elements

Sangeeta Sharma[1], Garima Bartariya[2*],
Shivali Rathore[3] and Juhie Agarwal[4]

1. Introduction

Rare Earth Elements (REEs) refer to a group of 17 chemically similar metallic elements that are critical to a wide range of modern technologies and industries (Srikant et al., 2021: Patel et al., 2024; Sehgal et al., 2025; Sharma et al., 2019). Despite their technological importance, the interaction of REEs with living organisms is a relatively underexplored area. Increasing evidence suggests that these elements can influence biological systems in diverse ways, ranging from stimulating growth and metabolism in plants and microbes to potential toxic effects depending on concentration and exposure conditions.

Biotechnology offers promising avenues to harness and study these interactions for various applications. For instance, certain microorganisms and plants have shown the ability to bioaccumulate or transform REEs, presenting opportunities for bioremediation of contaminated environments and bio-mining of these valuable metals. Furthermore, the unique properties of REEs are being explored to enhance agricultural productivity, develop biosensors, and improve drug delivery systems.

Living organisms need to control the selectivity of different molecules across their cell membrane to be able to access the resources that they need and manage molecules that could be harmful (De Rosa et al., 2023). There are conserved biological requirements for nutrient elements such as nitrogen, phosphorus and potassium, and many micronutrients, but it may be unexpected to find conserved requirements for REEs because they are not considered essential and excess REE accumulation is known to be toxic to living cells (Malhotra et al., 2020).

[1,2] Associate Professor, School of Life Science and Technology, IIMT University, Meerut, UP, India.
[3] Assistant Professor, Dept. of Botany, Govt. K.R.G. PG Autonomous College, Gwalior, MP, India.
[4] Assistant Professor, Vardhman PG College, Bijnor, UP, India.
* Corresponding author: garima_sls@iimtindia.net, dr.garimabartariya@gmail.com

DOI: 10.1201/9781003683254-7

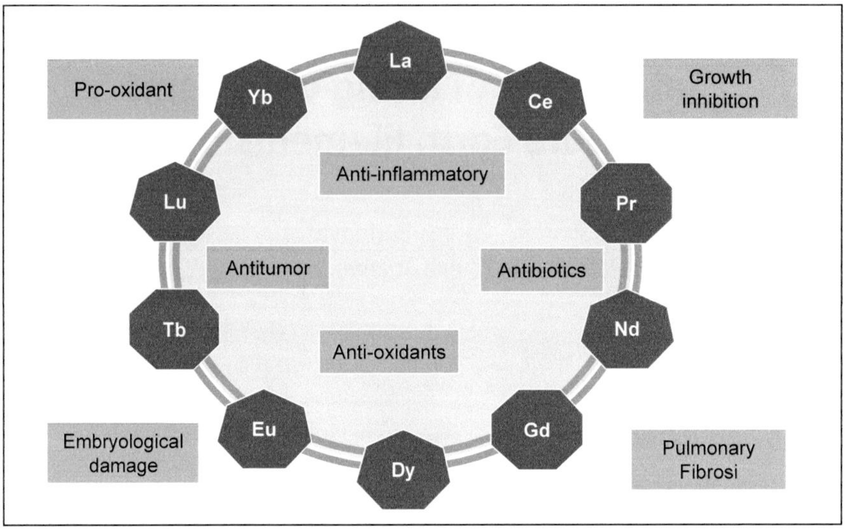

Fig. 1 Interaction of REEs with living organisms.

2. Mechanism of Interaction of Organisms with Rare Earth Elements

REEs are referred to as 'rare' while they are actually abundant in earth's crust (Sharma et al., 2019; Patel et al., 2025; Patel, et al., 2025; Patel et al., 2025). But despite their dispersal all over the crust, it is challenging to extract them. Interaction of these REEs with living organisms is a complex mechanism and can range from a lower to a higher concentration with beneficial to toxic effects simultaneously. It is the multifaceted interaction of these elements with livings that can involve their uptake mimicry of essential ions, induction of oxidative stress, direct protein binding, bioaccumulations, and detoxification strategies. In this interaction, various living things behave differentially. Some organisms adapt the need and utilize these REEs in special biochemical processes, while others cannot tolerate or adapt them and show detrimental effects on excessive exposure during various biological systems.

2.1 Breakdown of Mechanism of Interaction

2.1.1 Bioavailability and Speciation of Entry Routes

REEs take their pathway through natural and anthropogenic activities of the environment. These natural pathways include weathering of rocks majorly whereas anthropogenic activities include industrial waste, agricultural fertilizers, and mimicry, etc. Therefore, they can be available for organisms in soil, water, and air.

Table 1 Speciation of entry routes of REE.

Organism Type	*Entry Route*	*Key Factors Affecting Uptake*
Plants	Root uptake, foliar absorption	Soil pH, organic ligands, ion channels
Microorganisms	Surface adsorption, active transport	Surface charge, lanthanide-specific proteins
Animals	Ingestion, gill/skin absorption	Speciation, diet, environmental concentration

In plants, the REEs enter through the roots via the Apoplastic pathway **or** Symplastic pathway. Foliar Absorption also takes place in some plants from atmospheric deposition or foliar sprays (Shi et al., 2021). In microorganisms, REEs absorb through surface binding. REEs adsorb to cell wall functional groups (e.g., carboxyl, phosphate, hydroxyl). Some bacteria (e.g., *Methylacidiphilum fumariolicum*) possess lanthanide-specific ABC transporters to internalize REEs (Pol et al., 2014). In Aquatic Organisms, uptake occurs via gills or body surfaces or dietary uptake through contaminated food or sediment. In terrestrial animals, ingestion of REE-contaminated soil, plants, or water take place. In some animals, absorption occurs via gastrointestinal tract, influenced by speciation and co-occurring elements (Vassilova & Hoenig, 2001).

2.1.2 Cellular and Molecular Mechanism

Due to their extensive use in a variety of industries and possible effects on living things, REEs have attracted interest. Although they have historically been regarded as non-essential, new research has shed light on their intricate connections at the molecular and cellular levels in several biological systems.

Endocytosis Activation occurs via Arabinogalactan Proteins (AGPs). Lanthanum (La^{3+}) is one of the REEs that can interact with extracellular AGPs in plants to start clathrin-mediated endocytosis (CME). Usually tethered to the plasma membrane, these proteins alter structurally when La^{2+} is present, taking on a transmembrane form. The internalization of La^{3+} into the cell is the result of this conformational shift, which enables AGPs to engage with adaptor protein 2 (AP2) complexes (Wang et al., 2019). The cell wall is essential for the adsorption and detoxification of REEs in plants such as *Phytolacca americana*. Functional groups (-COOH and -OH) found in polysaccharides like hemicellulose and pectin bind REEs and lessen their toxicity. The cell wall's levels of galacturonic acid and total sugars rise in response to REE exposure, suggesting a mechanism to improve metal binding and tolerance (Guo et al., 2023).

Reactive oxygen species (ROS) produced by high REE concentrations can cause oxidative stress, which can change the flexibility of membranes and cause lipid peroxidation. For example, cerium (Ce^{4+}) disrupts cell membranes and interferes with water transport through aquaporins, making it more hazardous than its trivalent form (Ce^{3+}).

Microorganisms interact with REEs in a variety of ways, such as through biosorption, bioaccumulation, and bioleaching. While bioaccumulation involves intracellular accumulation and active transit, biosorption involves the passive adsorption of REEs to functional groups on the cell surface. Potential techniques for bioremediation of REE-contaminated environments are provided by the ability of certain bacteria to precipitate REEs as phosphate minerals through enzymatic activity (Vitova & Mezrieky, 2024).

2.1.3 Mechanism for Environmental Interaction

Because REEs are used so extensively in current technologies, agriculture, and industrial processes, they are becoming more and more common in the environment. Because of their distinct physicochemical characteristics, including their trivalent ionic state (except for Ce and Eu), ionic radii that are similar to calcium's, and propensity to form complexes, REEs interact with organisms in a complicated and diverse manner.

There is conflicting and limited data on the transferability of REEs along trophic chains. Accordingly, REEs that have been collected by lower trophic levels (plants, microbes, etc.) may be passed from herbivores to carnivores, potentially amplifying their effects farther up the food chain. REEs can have a detrimental effect on fertilization success, embryonic development, and reproductive performance in aquatic species like sea urchins, which can result in cytogenetic abnormalities and developmental problems (Sysolytina & Olkova, 2022).

There are some more examples of complexation within Organisms. Once inside an organism, REEs can be complexed by various organic molecules like proteins, organic acids, and phosphates. This complexation can help detoxify the REEs or facilitate their transport and storage. In some cases, REEs can be precipitated within cells (intracellular), for example, in vacuoles in fungi, as a detoxification mechanism

Microorganisms play a crucial role in the biogeochemical cycling of REEs. They can mobilize REEs by secreting organic acids or chelating agents that increase REE solubility. They immobilize REEs through biosorption or bioprecipitation, effectively removing them from the soluble phase. Microbial activity can also alter local pH and redox conditions, influencing REE speciation and mobility (Zhang et al., 2022).

3. Applications on Interaction of Living Organisms with Rare Earth Elements

3.1 Biomonitoring and Biosorption

REEs are found in ore as deposits and are complexed with phosphates, fluorides, and carbonates. These deposits occur as a result of microbial activities, environmental and tectonic factors. Microbial action includes dissolution, extraction, utilization, and mobilization of REEs in mineral deposits to alter the composition of solutions in the rocks. As in subtropical regions, regolith-hosted REE deposits are predominately found as a result of strong microbial activities. Recently it has been reported that the mobilization, distribution, and availability of REEs have been increased due to biological weathering, indicating the significant role of living organisms in the formation of REE deposits and its prevalence in the environment too (Li et al., 2022).

REEs exposure has been reported to act as an enzymatic co-factor, and enhance photosynthetic efficiency, biotic and abiotic stress response and adaptation in many plant species. Biological organisms can access REE and serve such elements as bioindicators or biomonitors as they respond to change the state of the environment by physiological activities resulting from pollution. In urban areas, mosses, lichens, mushrooms, many herbs and trees are used to get knowledge about the metal contamination due to its easy accessibility and high tolerance capacity (Isinkaralar et al., 2024). Studies revealed that numerous plants specially herbaceous plants are suitable well to be used as bioindicators and common dandelion (*Taraxacum officinale*), common nettle (*Urtica dioica*), and broadleaf plantain (*Plantago major*) have been reported to be used for environmental biomonitoring (Levei et al., 2018).

Dandelions exhibit a remarkable tolerance to various toxic substances, drought conditions, and nutritional deficiencies. Bivalve molluscs are frequently utilized as bioindicators of contamination due to their significant bioaccumulation capacity for a diverse array of contaminants, along with their continuous exposure to potentially contaminated organic and inorganic particles (Beyer et al., 2017). The concentrations found in the soft tissues of bivalves can serve as an indicator of the level of bioavailable contamination or pollution in a specific area.

Biosorption is a technique that involves the transfer of toxic metals from an aqueous environment to dead or living biomass.The green alga *Chlamydomonas reinhardtii* has been documented as a candidate for biosorption. Additionally, the carbonized biomass of non-living algae, specifically Parachlorella, has been utilized for the recovery of REEs from clay minerals (Ponou et al., 2014). Various green marine macroalgae, *including Cladophora fasicularis, Chaetomorpha sp, Caulerpa sertularioides, Ulva lactuca, and Valoniopsis pachynema* are also capable of accessing heavy metals (Kumar et al., 2009).

3.2 Bioimaging and Diagnostics

REEs are being utilized in various imaging methods, including computed tomography scans, positron emission tomography (PET) imaging, magnetic resonance imaging (MRI), and X-ray imaging REEs doped nanoparticles are being used in the fluorescence imaging as they allow to visualize cell activity and lesions. Such nanoparticles are water soluble and biocompatible to be used for bioimaging.

The Lanthanide series of REEs have been used remarkably in high-tech technologies like bioimaging and diagnostics as having distinctive physical, optical and magnetic properties, narrow luminescence emission spectra, etc. Lanthanide based luminescent bioassay and sensors have been reported to provide long lived emission. In molecular genetics Europium is being used to mark specific DNA strands in living tissues for luminescence by researchers (Nishioka et al., 2006). REEs as Gadolinium (Gd) can augment resolution and quality of any image thus, is being used in medical diagnostics like MRI as an effective contrasting agent. Gd (III) ions are paramagnetic compounds able to visualize the morphology of images with very high resolution by enhancing the magnetic relaxation of nuclei with low toxicity (Mitsumori et al., 2014). It does not need to be photobleached which makes it suitable for NIR (near- infrared bioimaging). Lanthanide-based nanoparticles or probes are extensively utilized due to their unique photophysical properties, which enable millisecond-scale luminescence with significant Stokes shifts, exceptionally sharp emissions, and robust resistance to photo-bleaching.

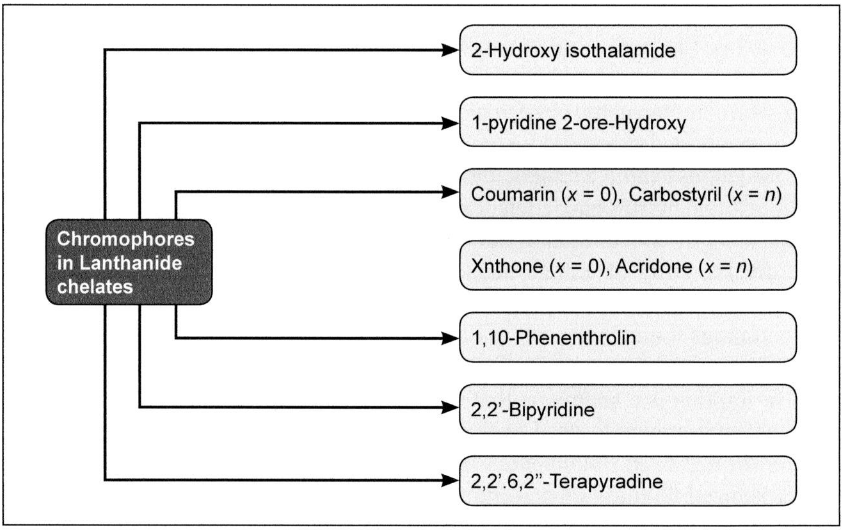

Fig. 2 Chromatophores of Lanthanide series.

Luminescence associated with lanthanide metal ions differs significantly from a parity-forbidden transition within the 4f atomic orbital, which imparts unique spectral characteristics to lanthanides, indicating a distinct photophysical process. Previous studies have examined the antenna effect and the ability of EDTA and NTA (nitrilotriacetate) chelates to mitigate aqueous quenching in lanthanide luminescence (Horrocks et al., 1977).

Biomedical imaging serves as a crucial instrument for visualizing specific tissues and comprehending biological processes at the molecular level. It is also beneficial for illustrating the internal structure of biological organisms, forming a foundation for future diagnoses or treatments. The MRI technique employs magnetic fields and radio waves to produce detailed images of the body's internal structures. Gadolinium, in its chelated form, is utilized as a contrast agent to enhance the visibility of specific tissues and organs during scans. However, recent research has reported evidence of gadolinium accumulation in neuronal tissues, which may pose risks to patients (McDonald et al., 1995).

Dual-modality molecular probes have been developed utilizing REEs to acquire more comprehensive diagnostic information. REEs have been particularly integrated into applications involving high-resolution MRI, Optical Imaging (OI), Radioisotope Imaging based on single-photon emission computed tomography (SPECT) and PET, as well as Computed Tomography (CT) and Ultrasonic Imaging (US), which provide highly sensitive and specific molecular probes. For real-time imaging, OI is employed due to its non-invasive nature, high sensitivity, and specificity; however, it suffers from limited light penetration through tissue (Rowe & Pomper, 2022).

Nanoparticle probes that incorporate REE offer adjustable surface properties, precise target delivery, and biocompatibility, rendering them effective for medical diagnosis and treatment. Gadolinium complexes (such as DOTA-Gd3+ and DTPA-Gd3+) are widely utilized for MRI probes due to their high stability and low biotoxicity (Do et al., 2020).

3.3 *Agricultural Promotion by REE*

Rare earth elements have been accumulated in soil by local geological parental rock. The concentration of these Elements (REEs) in soil varies significantly, ranging from 30 to 700 mg/kg, influenced by factors such as the type of parental rock, soil texture, organic matter content, pedogenic processes, and human activities. Currently, REEs are utilized as micro fertilizers and feed additives. Alongside macronutrients such as calcium, magnesium, nitrogen, phosphorus, and sulfur, micronutrients including iron, manganese, and silicon, as well as minerals and soil conditioners, also contain REEs and are widely employed. The ability to accumulate REEs from soil by plants varies as it depends on various factors like plant species, growth parameters, method of application and REE content in soil substrate (Fu et al., 2001). Plants accumulate such elements by roots or aerial parts. It has been reported that REE accumulation is higher in rice crops than corn.

Various forms of fertilizers containing REEs, such as nitrate, chloride, and ammonium bicarbonate, have been documented for commercial use. Typically, REEs are utilized either individually or in the forms of nitrate and chloride. Research has indicated that varying concentrations of REEs can have either beneficial or toxic effects on plants, leading to the recommendation that REEs should be applied to the aerial parts and seeds rather than directly to the soil. Agathokleous et al. (2019) noted that REEs induce hormesis in plants, where low concentrations have a stimulating effect while high concentrations can be toxic. At lower concentrations, REEs have been shown to positively influence the rate of photosynthesis by promoting chlorophyll synthesis and chloroplast development. Additionally, REEs enhance the activity of the rubisco enzyme and help mitigate various types of stress (Ma et al., 2022).

In *Vigna angularis*, lanthanum nitrate is utilized to enhance the efficiency of phosphorus assimilation. Certain lanthanide ions (La^{3+}, Nd^{3+}) may partially substitute for the endogenous calcium in plants and can also replace magnesium ions in chlorophyll. He and Loh (2000) investigated that the incorporation of Ce- or La nitrate during the vegetative stage has a beneficial impact on root development, although it does not affect plant biomass (He et al., 2000).

3.4 REEs as Enzyme Catalyst and Biosensors

Advanced oxidative processes have been developed to decontaminate waste water effectively. REEs has been used widely to catalyze this advanced oxidative process which degrade organic pollutants into CO_2, H_2O, and inorganic acids as they promote the formation of free radicals, such as hydroxyl (OH·), sulfate (SO42−·), and superoxide ion radicals (O2−·) (Spasiano et al., 2016). REEs possess distinct properties like high surface area and redox potential so that they can be used effectively in advanced oxidative processes.

Kajjumba et al. (2021) studied the use of lanthanides for waste water treatment that also represents an antagonistic effect thus reducing its impact on the environment. So there is a need to take care about the transfer of residual particles to the environment which is also based on some factors such as amount and quality of organic pollutants, pH, and concentration of cations. Such parameters also affect bioaccumulation of pollutants. Thus, alternative oxidative processes including REEs as catalysts must be adopted for environmentally safe waste water treatment by taking concern about the knowledge about the type of organic pollutants and acid base properties of waste water (Kajjumba et al., 2021)

In recent years, six rare earth oxides (REOs) as Ce_2O_3, Y_2O_3, Sm_2O3, Gd_2O_3, Dy_2O_2, and Yb_2O_3 possess multifarious applications in the field of biomedical science. In the biomedical field, development of biosensors has gained much attention which can be used for various aspects including pH sensors, ionic sensors and sensors to detect particular material in the target body. Studies revealed that Eu_2O_3 nanoparticles are being used to bind and catalyze certain bacteriophages in waste water. Sm_2O_2 is used as pH-sensitive membranes having high K value and its ability to anneal at a very high temperature. In order to develop an effective electrolyte-insulator-semiconductor (EIS) sensor, anions like chloride, nitrate, and Sm2O3 as an ion-sensing material are used.

3.5 Bioremediation of Rare Earth Elements Contamination

Bioremediation uses living organisms such as plants, animals, and microbes to break down or store environmental pollutants through natural processes. Soil bioremediation occurs through extraction, transformation, and immobilization of contaminants. Although immobilization does not change the concentration of contaminants, it lowers the environmental risk by changing how easily these contaminants move and are available for living things. Bioremediation mainly occurs through natural methods, causing little disruption to the environment. For instance, it relies on natural processes, making it a cost-effective and eco-friendly option for cleaning soil. Pohlia flexuosa moss is particularly good at absorbing La to extract REEs from polluted areas while also preserving well. Bioremediation usually involves cleaning contaminated soil on-site, which can be slow and heavily relies on environmental conditions like climate and the type of soil at the cleanup location (Schmoger et al., 2000). Human mining activities inevitably spread metal elements, with REEs becoming important pollutants. Therefore, methods such as using plants, microbes, or a combination of both can be used to remove contaminants from REE-contaminated soil.

3.5.1 Phytoremediation of REE-Contaminated Soil

Phytoremediation is a primary bioremediation technique that employs plants to extract pollutants from the soil sustainably through processes such as phytoextraction, rhizofiltration, and phytodegradation. It also improves plant growth, restores ecosystems in mining areas, and enhances the landscape's appearance. Additionally, it allows for the recovery of REE resources using special plants in phytomining systems. The effectiveness of this process is also influenced by soil and weather conditions that affect plant health. Soils contaminated with rare-earth metals can be cleaned up using different phytoremediation methods (Sharma et al., 2023). Among these, phytoextraction and phytostabilization are the most common methods for soils polluted with potentially harmful elements, and they can also be used for soils with REEs. Phytoextraction is a simple and low-cost method that involves growing and harvesting hyper-accumulator plants that take in high amounts of metals in their shoots, helping to eliminate metals from contaminated soil. Although high levels of potentially toxic elements can harm plant health and cause other negative effects, hyper-accumulator plants can tolerate and accumulate these elements in their living shoots. In contrast, phytostabilization helps keep pollutants from moving in the soil, allowing plants to absorb them through their roots or trap them in the root zone.

Plant species and the particular metal have a significant impact on the distribution of REEs within plants. The concentration of REEs in the subterranean portions of plants is typically higher than in the above-ground portions. Potentially harmful effects of elevated lanthanum (La) levels include disruptions to cell division, DNA structure, nutrition absorption, and photosynthesis. Through detoxifying methods include creating certain chemicals, storing it in vacuoles, and using defense mechanisms against oxidative stress, plants may absorb a limited quantity of La.

Accurately measuring REE levels in the organ systems of vascular plants is still difficult due to issues with identifying particles associated with their roots. For example, vegetables like cabbage have very low REE levels. The way REEs are spread in the roots, stems, and leaves of vascular plants varies greatly. Comparing different plant parts shows that more REEs are often found in roots, likely due to limited movement capabilities rather than issues with transferring them to the above-ground parts. Studies have shown that La gathers differently in root and shoot tissues of crops like soybeans, corn, and mung beans, with roots keeping more of this element.

3.5.2 Remediation of REE-Polluted Soil by Microorganisms

Microbial cleanup is carried out through processes used by microbes to change harmful substances in soil. This method allows for on-site treatment of soil, removing the need for digging and lowering the risk of spreading pollutants during transport. It helps break down organic matter and promotes nutrient cycling, which improves soil health. Groups of microbes show the ability to break down stubborn organic pollutants and hold onto heavy metals, handling mixed contaminants and being useful in many different soil types.

The effectiveness of microbial cleanup is limited by natural genetic differences and the tendency for mutations in these microbial groups. Slower metabolic rates mean that treatment periods often need to be longer, usually achieving only partial removal of contaminants. The success of cleanup is also influenced by soil conditions like temperature, moisture levels, and changes in pH, and oxygen availability, all of which are important for achieving the best results. Microbes can tolerate heavy metals but have a specific range; extreme changes can harm their growth, making it harder for them to absorb and hold heavy metals. Improving bioremediation in contaminated soils can be done by adjusting the soil environment to boost microbial life. This strategy requires adding extra nutrients to improve the availability of nutrients in the soil, which increases microbial growth and activity, thus enhancing the breakdown of contaminants. Effective cleanup organisms usually have small cell sizes, fast growth rates, strong metabolic processes, and the ability to adapt to different environments. These characteristics are especially helpful in cleaning REE-contaminated soils.

Activities in the rhizosphere and microorganisms help to produce oxalate species, which then form complexes with REEs under slightly acidic environments, enhancing the interactions between REE and phosphate minerals. Actinobacteria efficiently leach REEs from bastaseite materials by producing organic acids and siderophores, which aid in the transport of REEs. Microalgae can be used to recover REEs from a variety of sources, such as industrial wastewater, electronic waste, and ores. Numerous studies have examined the biosorption of lanthanum (La), showing promise in Pseudomonas and Myxococcus xanthus. Under comparable experimental conditions, comparisons show that biofilms have a 160% higher capacity to absorb La than the *Euglena mutabilis* suspension.

The combined cleanup method using plants and microorganisms shows great benefits through a partnership between roots, microbes, and pollutants; specifically, root secretions and microbial products work together to increase the availability of REEs, while mycorrhizal partnerships help to hold contaminants in place and improve nutrient uptake

4. Bioremediation of REE-Polluted Soil Using the Plant–Microorganism Combination

Combined plant and microorganism cleaning systems improve the removal of pollutants better than single method approaches, helping speed up the cleaning of sites through working together in biological processes. Plant growth-promoting bacteria (PGPR) can help move metals and produce plant hormones, thus boosting the effectiveness of REE extraction by making nutrients more available to plants. Mycorrhizal partnerships improve soil health by increasing nutrient and water uptake, stabilizing soil conditions, and helping plants resist biological and environmental stresses (Jalaliu & Lebeau, 2021). Fungal substances like hartzianic acid can both encourage plant growth and have antibiotic properties. Groups of microbes improve phytoremediation by making pollutants more available and helping plants handle stress through secondary substances (Alsherif et al., 2024). Arbuscular mycorrhizal fungi (AMF) can form helpful relationships with over 80% of land plants, reducing the harmful effects of La while enhancing the plant's ability to absorb water and nutrients.

The colonization by AMF has enhanced the absorption of vital minerals such as calcium (Ca), potassium (K), zinc (Zn), and nitrogen (N) under both normal conditions and those stressed by europium (Eu). Additionally, it has mitigated oxidative damage induced by Eu, likely through the improved activation of protective nutrient systems. This is particularly evident in the movement of metals facilitated by PGPR, the nutrient absorption capabilities of AMF, and the role of fungal metabolites in alleviating stress. Nevertheless, while the interactions between REEs, soil microorganisms, and plant roots are crucial for maintaining soil health and enhancing cleaning efficacy, further research is essential to gain a comprehensive understanding of these dynamics. By merging various disciplines and employing innovative technologies, the combined plant-microorganism cleanup approach presents a promising strategy for addressing soil contaminated with REEs, thereby fostering advancements in ecological restoration and the sustainable utilization of resources in regions impacted by REE pollution.

5. Bioextraction and Biorecovery of Rare Earth Elements

REEs are becoming more and more necessary on a global scale (Sharma et al., 2023; Patel et al., 2024; Sharma et al., 2021). They are essential to many devices, from renewable energy systems to cell phones. Magnets, which are essential for electronic cars and wind turbines (Sharma et al., 2019) are made from REEs like neodymium (Nd), praseodymium (Pr), terbium (Tb), and dysprosium (Dy) (Sharma et al., 2019; Sharma et al., 2022). Most REEs are extracted using hazardous acids and solvents after being mined from limited ore sources.

Large volumes of hazardous trash can be produced by chemical methods of recovering REEs from urban waste. Nevertheless, although they have not yet been used commercially, biological techniques can lessen hazardous waste and negative environmental effects. Numerous methods have been thoroughly investigated for the extraction of valuable metals in the mining and wastewater treatment sectors.

The method of obtaining REEs via the enzymatic actions of microorganisms or their metabolic byproducts is referred to as bioleaching. Using bioleaching, REEs can be efficiently and cost-effectively extracted from waste materials and low-grade ores.

Reaction times are frequently lengthy, and maintaining ideal microbial growth conditions is necessary to ensure consistent output and yield—these are the main drawbacks of bioleaching. Microbial bioleaching has been used in a number of studies to recover REEs from ores, industrial mining wastes, and e-waste. Biosorption is a process that uses biomass or biological materials to attach REEs to their cellular framework. In addition to producing little waste and having a high process efficiency, this approach can also be economical and enable sorbent replenishment. Various biomass types, including bacteria, algae, fungi, and plants, were used in the first biosorption studies to extract REEs from wastewater. Hosomomi et al. (2013) looked into the adsorption properties of REEs on Escherichia coli. They concluded that E. coli exhibited optimal performance in a solution with a higher pH and demonstrated greater selectivity for heavy REEs (Hosomomi et al., 2013).

REEs and heavy metals can be precipitated via a process called bioprecipitation, which employs metabolites produced by microorganisms. Bacteria were employed in early bioprecipitation investigations to extract heavy metals, including cadmium (Cd). Alkaline phosphatase's enzymatic degradation of p-nitrophenyl phosphate as a source of inorganic phosphate for precipitation was investigated by some researchers (Chaudhuri et al., 2013). When they tested their new method for removing heavy metals from industrial waste, they discovered that precipitation was an efficient way to remove Cr^{6+}, Cr^{3+}, Ni^{2+}, Cd^{2+}, and Co^{2+}.

Ionic liquids have shown superior performance compared to conventional solvents and offer sustainability benefits such as waste minimization and pollution prevention, while organic solutions have been used for solvent extraction for an extended period.

6. Conclusion

REEs are not essential elements for most organisms, so their biological roles are not well-understood.Emerging research shows that some microorganisms can utilize REEs (e.g., *methylotrophic bacteria* that use lanthanides in methanol dehydrogenase), but the mechanisms remain poorly characterized. Many microbes have limited tolerance to REE concentrations typically found in industrial waste or ores. Engineering microbes for selectivity and tolerance is an ongoing challenge in synthetic biology.

Genetically engineered microbes or plants can be designed to enhance REE uptake, selectivity, and resistance. Pathways involved in REE-binding proteins or transporters can be manipulated to improve bioaccumulation or bio-mining efficiency. Certain plants (hyperaccumulators) may be used to extract REEs from contaminated soils (phytomining). Discovery and design of peptides or proteins that bind REEs with high specificity could enable biosensors, filters, or separation technologies. These could be integrated into bioreactors for recovery from e-waste, water, or industrial effluents. Understanding the ecological fate and long-term impact of REEs is crucial for sustainable development. Biotechnology can play a key role in extracting REEs from electronic waste through eco-friendly processes. Integrating bio-based solutions into circular economy models offers promising opportunities for sustainability.

References

Agathokleous, E., Kitao, M. and Calabrese, E.J. (2019). Hormetic dose responses induced by lanthanum in plants. *Environmental Pollution*, 244, 332–341.

Alsherif, E.A., Sonbol, H., AbdElgawad, H., Ramadan, A., Korany, S.M., Crecchio, C. and Bouqellah, N.A. (2024). Arbuscular mycorrhizal fungi improve tolerance of wheat plants under soil Europium contamination. *Plant and Soil*, 1–15.

Beyer, J., Green, N.W., Brooks, S., Allan, I.J., Ruus, A., Gomes, T. and Schøyen, M. (2017). Blue mussels (Mytilus edulis spp.) as sentinel organisms in coastal pollution monitoring: A review. *Marine Environmental Research*, 130, 338–365.

Chaudhuri, G., Shah, G.A., Dey, P., S, G., Venu-Babu, P. and Thilagaraj, W.R. (2013). Enzymatically mediated bioprecipitation of heavy metals from industrial wastes and single ion solutions by mammalian alkaline phosphatase. *Journal of Environmental Science and Health, Part A*, 48(1), 79–85.

De Rosa, A., McGaughey, S., Magrath, I. and Byrt, C. (2023). Molecular membrane separation: Plants inspire new technologies. *New Phytologist*, 238(1), 33–54.

Do, C., DeAguero, J., Brearley, A., Trejo, X., Howard, T., Escobar, G.P. and Wagner, B. (2020). Gadolinium-based contrast agent use, their safety, and practice evolution. *Kidney360*, 1(6), 561–568.

Fu, F., Akagi, T., Yabuki, S. and Iwaki, M. (2001). The variation of REE (rare earth elements) patterns in soil-grown plants: A new proxy for the source of rare earth elements and silicon in plants. *Plant and Soil*, 235, 53–64.

Guo, Y., Chen, K., Lei, S., Gao, Y., Yan, S. and Yuan, M. (2023). Rare earth elements (REEs) adsorption and detoxification mechanisms in cell wall polysaccharides of Phytolacca americana L. *Plants*, 12(10), 1981.

He YaWen, H.Y. and Loh ChiangShiong, L.C. (2000). Cerium and lanthanum promote floral initiation and reproductive growth of Arabidopsis thaliana. *Plant Science*, 159(1), 117–124.

Horrocks Jr, W. D., Schmidt, G.F., Sudnick, D.R., Kittrell, C. and Bernheim, R.A. (1977). Laser-induced lanthanide ion luminescence lifetime measurements by direct excitation of metal ion levels. A new class of structural probe for calcium-binding proteins and nucleic acids. *Journal of the American Chemical Society*, 99(7), 2378–2380.

Hosomomi, Y., Baba, Y., Kubota, F., Kamiya, N. and Goto, M. (2013). Biosorption of rare earth elements by Escherichia coli. *Journal of Chemical Engineering of Japan*, 46(7), 450–454.

Isinkaralar, O., Świsłowski, P., Isinkaralar, K. and Rajfur, M. (2024). Moss as a passive biomonitoring tool for the atmospheric deposition and spatial distribution pattern of toxic metals in an industrial city. *Environmental Monitoring and Assessment*, 196(6), 513.

Jalali, J. and Lebeau, T. (2021). The role of microorganisms in mobilization and phytoextraction of rare earth elements: A review. *Frontiers in Environmental Science*, 9, 688430.

Kajjumba, G.W., Attene-Ramos, M. and Marti, E.J. (2021). Toxicity of lanthanide coagulants assessed using four in vitro bioassays. *Science of the Total Environment*, 800, 149556.

Kumar, J.N., Oommen, C. and Kumar, R.N. (2009). Biosorption of heavy metals from aqueous solution by green marine macroalgae from Okha Port, Gulf of Kutch, India. *American-Eurasian J. Agric. and Environ. Sci.*, 6(3), 317–323.

Levei, L., Kovacs, E., Hoaghia, M.A. and Ozunu, A. (2018). Accumulation of heavy metals in Plantago major grown in urban and post-industrial areas. *Studia Ubb Chemia*, 63(1), 87–98.

Li, X., Liang, X., He, H., Li, J., Ma, L., Tan, W. and Dong, H. (2022). Microorganisms accelerate REE mineralization in supergene environments. *Applied and Environmental Microbiology*, 88(13), e00632–22.

Ma, Y., Wang, T., Xie, Y., Lv, Q. and Qiu, L. (2022). Alleviatory effect of rare earth micro-fertilizer on photosystem II (PSII) photoinhibition in Pseudostellaria heterophylla leaves at photosynthetic midday depression. *Journal of Rare Earths*, 40(7), 1156–1164.

Malhotra, N., Hsu, H.S., Liang, S.T., Roldan, M.J.M., Lee, J.S., Ger, T.R. and Hsiao, C.D. (2020). An updated review of toxicity effect of the rare earth elements (REEs) on aquatic organisms. *Animals*, 10(9), 1663.

McDonald, J.W., Ghio, A.J., Sheehan, C.E., Bernhardt, P.F. and Roggli, V.L. (1995). Rare earth (cerium oxide) pneumoconiosis: Analytical scanning electron microscopy and literature review. *Modern Pathology: An Official Journal of the United States and Canadian Academy of Pathology, Inc*, 8(8), 859–865.

Mitsumori, L.M., Bhargava, P., Essig, M. and Maki, J.H. (2014). Magnetic resonance imaging using gadolinium-based contrast agents. *Topics in Magnetic Resonance Imaging*, 23(1), 51–69.

Nishioka, T., Yuan, J., Yamamoto, Y., Sumitomo, K., Wang, Z., Hashino, K. and Matsumoto, K. (2006). New luminescent europium (III) chelates for DNA labeling. *Inorganic Chemistry*, 45(10), 4088–4096.

Patel, D.K., Sharma V.K. and Sharma H.O. (2024). Advancements in aluminum matrix composites: A comprehensive review on the mechanical and microstructural attributes of rare earth elements (REEs) and rare earth oxides (REOs) doped materials. *Transactions of the Indian Institute of Metals*, 77(9), 2265–2276.

Patel, D.K., Sharma, V.K. Sharma, H.O. and Kumar, P. (2024). Application of AI in Material Science to Accelerate Material Innovation. In: *Artificial Intelligence in Material Science*, pp. 198–207. CRC Press.

Patel, D.K., Sharma, V.K., and Sharma. H.O. (2025). Current global Status of rare earth elements (REEs) and their role as catalysts in reducing air pollution for maintaining environmental sustainability. *Transactions of the Indian Institute of Metals*, 78(1) 4.

Patel, D.K. and Sharma V.K. (2025). Illuminating Innovations: A Comprehensive Review of Rare Earth-Doped Phosphors and Their Applications in LEDs. *Transactions of the Indian Institute of Metals*, 78(7), 165.

Patel, D.K., Sharma, V.K., Kumar V., Kumar, P. and Kumar, D. (2025). Enhancing hydrophobicity and anti-corrosion properties of Al-6061Aluminum hybrid composites through trace additions of rare earth oxides. In: *Proceedings of the Institution of Mechanical Engineers, Part E: Journal of Process Mechanical Engineering*, 09544089251317381.

Pol, A., Barends, T.R., Dietl, A., Khadem, A.F., Eygensteyn, J., Jetten, M.S. and Op den Camp, H.J. (2014). Rare earth metals are essential for methanotrophic life in volcanic mudpots. *Environmental microbiology*, 16(1), 255–264.

Ponou, J., Wang, L.P., Dodbiba, G., Okaya, K., Fujita, T., Mitsuhashi, K. and Noda, M. (2014). Recovery of rare earth elements from aqueous solution obtained from Vietnamese clay minerals using dried and carbonized parachlorella. *Journal of Environmental Chemical Engineering*, 2(2), 1070–1081.

Rowe, S.P. and Pomper, M.G. (2022). Molecular imaging in oncology: Current impact and future directions. *CA: A Cancer Journal for Clinicians*, 72(4), 333–352.

Schmoger, M.E., Oven, M. and Grill, E. (2000). Detoxification of arsenic by phytochelatins in plants. *Plant physiology*, 122(3), 793–802.

Sehgal, A., Sharma, V.K., Gautam, P.S. and Partap A. (2025): Forging the Future: Exploring Aluminium AA-6061 Composites through Stir Casting for Advanced Applications and Future Innovations. *Evergreen*, 12(2), 806–826.

Sharma, V.K., Sharma, D.K. and Sharma S. (2023). Bio-Mimetic and Bio-Templating of Natural Materials: An Approach to Sustainable Development Using Bionanotechnology. In: *Bionanotechnology towards Green Energy*, pp. 187–206. CRC Press.

Sharma, V.K., Kumar, V. and Joshi R.S. (2019). Investigation of rare earth particulate on tribological and mechanical properties of Al-6061 alloy composites for aerospace application. *Journal of Materials Research and Technology*, 8(4), 3504–3516.

Sharma, V.K., Kumar, V., and Joshi, R.S. (2019). Effect of RE addition on wear behavior of an Al-6061 based hybrid composite. *Wear*, 426, 961–974.

Sharma, V.K., Kumar, V. and Joshi, R.S. (2021). Manufacturing of stable hydrophobic surface on rare-earth oxides aluminium hybrid composite. In: *Proceedings of the Institution of Mechanical Engineers, Part E: Journal of Process Mechanical Engineering*, 235(4), 899–912.

Sharma, V.K., and Kumar, V. (2019). Study on material transfer and surface properties during fiber laser cutting of A653 galvanized steel sheet. *Journal of the Brazilian Society of Mechanical Sciences and Engineering*, 41(8), 41.

Sharma, M., Janardhan, G., Sharma, V.K., Kumar, V. and Joshi R.S. (2022). Comparative prediction of surface roughness for MAFM finished aluminium/silicon carbide/aluminium trioxide/rare earth oxides (Al/SiC/Al2O3)/REOs) composites using a Levenberg–Marquardt Algorithm and a Box–Behnken Design. In: *Proceedings of the Institution of Mechanical Engineers, Part E: Journal of Process Mechanical Engineering*, 236(3), 790–804.

Sharma, J.K., Kumar, N., Singh, N.P. and Santal, A.R. (2023). Phytoremediation technologies and their mechanism for removal of heavy metal from contaminated soil: An approach for a sustainable environment. *Frontiers in Plant Science*, 14, 1076876.

Shi, K., Liu, C., Liu, D., Lyu, K., Chen, J. and Wang, X. (2021). The accumulation and effect of rare earth element neodymium on the root of rice seedlings. *Environmental Science and Pollution Research*, 28(35), 48656–48665.

Spasiano, D., Siciliano, A., Race, M., Marotta, R., Guida, M., Andreozzi, R. and Pirozzi, F. (2016). Biodegradation, ecotoxicity, and UV254/H2O2 treatment of imidazole, 1-methyl-imidazole and N, N'-alkyl-imidazolium chlorides in water. *Water Research*, 106, 450–460.

Srikant, Sharma, V.K., Kumar, V., Joshi, R.S., Jain, S. and Kumar, P. (2021). A review of recent research on rare earth particulate composite materials and structures with their applications. *Transactions of the Indian Institute of Metals*, 74(11), 2569–2581.

Sysolyatina, M.A. and Olkova, A.S. (2022). Sources of rare earth elements in the environment and their impact on living organisms. *Environmental Reviews*, 31(2), 206–217.

Vassileva, E. and Hoenig, M. (2001). Determination of arsenic in plant samples by inductively coupled plasma atomic emission spectrometry with ultrasonic nebulization: A complex problem. *Spectrochimica Acta Part B: Atomic Spectroscopy*, 56(2), 223–232.

Vítová, M. and Mezricky, D. (2024). Microbial recovery of rare earth elements from various waste sources: A mini review with emphasis on microalgae. *World Journal of Microbiology and Biotechnology*, 40(6), 189.

Wang, L., Cheng, M., Yang, Q., Li, J., Wang, X., Zhou, Q. and Deng, X. W. (2019). Arabinogalactan protein–rare earth element complexes activate plant endocytosis. In: *Proceedings of the National Academy of Sciences*, 116(28), 14349–14357.

Zhang, X., Hu, Z., Pan, H., Bai, Y., Hu, Y. and Jin, S. (2022). Effects of rare earth elements on bacteria in rhizosphere, root, phyllosphere, and leaf of soil–rice ecosystem. *Scientific Reports*, 12(1), 2089.

8 Advanced Applications of Rare Earths Materials in Nuclear Power Plant

Santosh Kumar Rai[1], Pardeep Kumar[2], Dhowmya Bhatt[3], Vikas Goyat[4*], Vipin Kumar Sharma[5], Gourav Kumar Sharma[6], Mahesh Kumar Gupta[7] and Shallender Bhati[8]

1. Introduction

Rare earth materials (REMs), also known as rare earth elements (REEs) comprise seventeen chemical elements, such as lanthanides, scandium, and yttrium, etc., are shown in Figure 1. They have attracted various researchers to carry out research in this field due to thermo, mechanical, and physicochemical properties (Srikant et al., 2021; Patel et al., 2024; Sharma et al., 2020a; Sharma et al., 2020b; Patel et al., 2025a; Patel et al., 2025b; Sharma et al., 2022). Recently, the use of REMs has been expanded for employing in different advanced nuclear reactors and nuclear power plants (NPPs) to increase the performance and safety; they are also used in advancing the NPPs due to high thermal stability, neutron absorption ability, and magnetic properties (Henderson, 2013). Due to these properties of the REMs, they are used in various parts of the NPPs such as control rods, fuel pallet, neutron absorber, burnable poisons, radiation shield, and advanced fuel designs. Some REMs such as gadolinium and samarium are used as neutron absorbers in control rods and radioactive materials due to their high neutron capture ability that helps regulate reactor reactivity and improve the operational safety of the NPPs (Bessais, 2021). REMs are added in the fuel of the NPPs to improve fuel performance and durability at higher temperature and pressure of the reactor, as well as application of REMs in the NPPs are also used for studying the development of accident-tolerant fuels and next-generation nuclear reactors, such as boiling water reactors, supercritical water reactors (SCWR) (Gupta & Krishnamurthy, 1992), supercritical

[1,4,7] Department of Mechanical Engineering, Faculty of Engineering and Technology, SRM Institute of Science and Technology, Delhi NCR Campus, Ghaziabad, Uttar Pradesh, India (201204).

[2] Department of Mechanical Engineering, Maharishi Markandeshwar (Deemed to be University), Mullana, Haryana, India, (133207).

[3] Department of Computer Science Engineering, Faculty of Engineering and Technology, SRM Institute of Science and Technology, Delhi NCR Campus, Ghaziabad, UP, India (201204).

[5,8] Department of Mechanical Engineering, Meerut Institute of Engineering and Technology, Meerut, U.P., 250001, India.

[6] Department of Mechanical Engineering, NIT, Jalandhar, Punjab, 144011, India.

* Corresponding author: viky.goyat@gmail.com

DOI: 10.1201/9781003683254-8

natural circulation loop (SCNCL) (Upadhyay et al., 2018; Rai et al., 2021a; Rai et al., 2020; Rai et al., 2021b), etc. Also, due to the growing demands for sustainable and safe nuclear energy with advancement of NPPs, REMs play a very critical role for improving the efficiency, safety, and life of the NPPs.

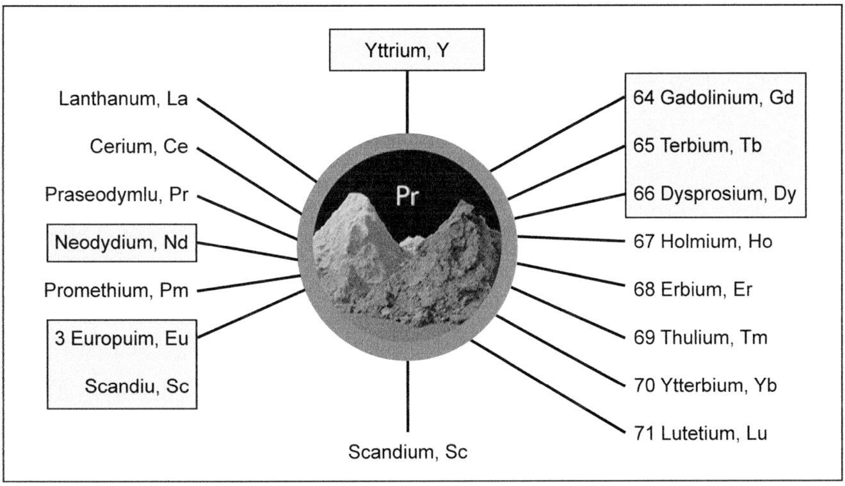

Fig. 1 Classification of REMs.

REMs are very important for the advancement application and functioning of SCNCLs and SCWRs because of their high thermal, structural, and neutron absorption characteristics (Rai et al., 2017); In SCNCLs, which are used for removing heat from the core as well as operating at supercritical pressure and temperature to produce electricity (Rai et al., 2025; Rai et al., 2022; Rai et al., 2020) consist of various parts such as electro-mechanical parts, sensor, thermocouple, etc., which require higher thermal stability properties that can be improved by adding rare earth oxides such as Yttria and Ceria in high-temperature sensor components and thermal barrier coatings due to their stability and corrosion resistance in supercritical pressure and temperature (Rai & Dutta, 2018; Idemitsu et al., 2003). Similarly, SCWRs operate at supercritical pressures which consist of radioactive materials in fuel rods and dissipate a huge amount of heat to the fluid during the fission reaction. Thus, REMs such as gadolinium (Gd) are added in the fuel which works as reactive materials in fuel assemblies and also as a moderator to reduce excess reactivity and improve the safety of the NPPs. Additionally, these materials are also used in the cladding and in the structural components due to their capacity to prevent radiation and thermal stress. By the addition of REMs in the NPPs, we can improve the efficiency and safety as well as increase the thermal stability of the structure at extreme operating conditions such as higher temperature and pressure (Elyassi et al., 2004).

The advancement and safety of NPPs depend on REMs, which have higher mechanical and chemical properties. Nowadays, REMs are used in different parts of NPPs to improve the performance and safety of the plants due to their high melting points, thermal stability, and neutron absorption capacity. Several parts of nuclear reactors consist of REMs such as gadolinium, dysprosium, samarium, and cerium for absorbing neutrons during fission reaction in the fuel. They are also used with the control rod to reduce burnable poison materials and control the nuclear reactivity and act like a moderator. Similarly, dysprosium and samarium help to increase reactor control and stability's functioning. REMs oxides and high-temperature ceramics are also used in NPPs as radiation shields for preventing radiation and working as coatings over the wall of NPPs to increase the life and performance of nuclear components. REMs are used to increase the safety, efficiency, and spam-life of both the present and next-generation nuclear plants. In this work, various applications of REMs in NPPs are discussed and their roles in fuel design, reactor moderation, waste management, and advanced material development is emphasized..

2. Importance of Rare Earth Materials in Nuclear Reactors

REMs are used in various NPPs and advanced nuclear reactors due to its higher thermal, mechanical, and neutron absorption properties. To enhance reactor performance and safety, various REMs are added in the different parts of the NPPs such as structural materials, control rods, radiation shielding, and fuel elements to enhance the efficiency of the NPPs dependent on fuel. They have the ability to stabilize fuel, absorb neutrons, and withstand extreme conditions (Zhao et al., 2023; Chidambaram and Banerjee, 2003).

2.1 Roles of REMs in Nuclear Power Plant Components

By mixing REMs in the nuclear fuels increases their melting points and mechanical properties. For example, yttrium and lanthanum act as stabilizers in actinide-rich fuels, including uranium and plutonium mixtures, to improve their performance in high-temperature and radiation environments. The incorporation of these elements diminishes the neutron penalty, facilitating enhanced fuel combustion and enabling the design of more compact reactor cores. Furthermore, the irradiation of REEs results in the formation of short-lived radioisotopes, thereby facilitating the process of nuclear waste disposal (Zhao et al., 2023). REMs such as dysprosium titanates, hafnates, and zirconates, which are used as neutron-absorbing materials in the fuel of NPPs due to their phase change properties, react with neutrons and have higher radiation resistance capacity (Kim et al., 2015). These REMs are mainly used in control rods and blankets to manage neutron flow and stop too many reactions, and their ability to work well with building materials and resist rust makes them valuable in NPPs. Similarly, REM oxides are very useful in preventing radiation; therefore, they are used in NPPs as a radiation shield to prevent the radiation from the core from reaching the plants. Radioactive rays are not capable of crossing the shield due to its higher density of REMs oxide as well as its properties to absorb gamma and neutron radiation. Europium and gadolinium oxides are mainly used in shielding glasses to improve the radiation absorption capabilities of the shield (Guon et al., 1990). Further, these REM oxides are used in dosimetry and radiation sensors to increase their sensitivity to accurate measurements of radiation exposure due to their optical and chemical properties. Also, the structure of the NPPs and advanced nuclear reactors are generally made of high-entropy alloys (HEAs) such as Cantor alloy, Y-containing HEAs, Gd-containing HEAs, etc., which consist of REMs. These alloys have higher radiation resistance and corrosion capability as well as the ability to withstand higher temperatures than Gd-containing HEAs. Further, REMs-based ceramic composites, such as SiC and Si3N4, are used in the development of the core of NPPs and GIV-IV nuclear reactors due to their high-temperature sustainable properties and higher radiation-preventing properties (Chernov et al., 2025; Hossain et al., 2022; Yvon et al., 2015).

2.2 Improving Thermal Efficiency, Performance, and Safety of NPPS using REMs

There are various ways to improve the thermal performance and efficiency of NPPs. However, many methods typically focus on efficiency enhancement without considering the NPPs' safety. Therefore, REMs are used to improve the efficiency and provide better safety of the NPPs. The following parts of the NPPs are improved by adding REEs in the nuclear reactor, which are given below. By adding REEs, such as rare earth stabilizers, in the nuclear reactor fuel, the neutron economy which helps improve the fuel cycle performance as well as the fission reaction to produce higher energy. Furthermore, this kind of fuel can be used in various compact NPPs to produce high energy (Guon et al., 1990). Similarly, by adding REEs such as dysprosium compounds, which work as neutron absorbers, the number of neutrons in the fuel increases. This outcome has shown the higher energy output and reduced fuel reprocessing requirements as well as increasing higher control of neutron flux and more efficient reactor operation (Kim et al., 2015; Yvon et al., 2015). REMs are basically higher entropy alloys such as ceramic composites, which can be sustained at higher temperatures and allow NPPs to operate at higher temperatures to reach maximum thermal efficiency while maintaining the safety of the NPPs (Igunma et al., 2024). Additionally, incorporating REEs into nuclear fuels enhances the fuel's ability to absorb neutrons, improves neutron economy, and reduces the production of long-lived radioisotopes. This results in increased nuclear energy waste and shorter disposal periods (Zhao et al., 2023; Guon et al., 1990; Guo et al., 2024; Watkins, 2015). Furthermore, by adding REMs in the NPPs in different parts reduces the global warming and adverse effects of radioactive fuel materials on the environment as well as it increases energy conversion efficiency in compact designs of NPPs to achieve higher thermal efficiency and performance, which reduces the overall environmental impact of nuclear power generation (Guon et al., 1990; Hossain et al., 2022; Kim et al., 2015; Chernov et al., 2025; Hossain et al., 2022; Yvon et al., 2015; Qiao et al., 2018; Igunma et al., 2024; Guo, 2024; Watkins, 2015).

2.3 Different Types of NPPSs Using REMs

REMs are widely used in nuclear reactors due to their exceptional properties, such as thermal stability, corrosion resistance, neutron absorption, and waste management capabilities. There are four types of reactors currently in use, and all of them utilize REMs for different applications. In fast reactors, REMs like lanthanum and yttrium are mixed with fuel to improve mechanical properties and increase the melting point of fuel rods (Zhao et al., 2023). REMs such as dysprosium titanate are also used in control rods and blankets to absorb neutrons and maintain neutron flux (Kim et al., 2015; Chernov et al., 2025; Hossain et al., 2022; Yvon et al., 2015; Qiao et al., 2018; Igunma et al., 2024; Guo, 2024; Watkins, 2015). In pressurized water reactors (PWRs), gadolinium and samarium oxides are utilized for neutron absorption to ensure safer and more efficient reactivity control (Qiao et al., 2018). Molten salt reactors (MSRs) use cerium and yttrium oxide coatings on structural components to protect them from corrosion and radiation damage (Qiao et al., 2018; Igunma et al., 2024; Guo, 2024; Watkins, 2015; Malerba et al., 2022). In modern Generation-IV reactors, REMs like lanthanum, yttrium, and cerium-based high-entropy alloys, and ceramics are employed to enhance thermal stability and radiation resistance (Ballaud et al., 2021; Niarchos 2021; Lee et al., 2013). Furthermore, gadolinium and dysprosium are used to control neutron flux (Marra, 2011). Further, various applications of the different REMs in terms of cost is given in Figure 2 which depicts the percentage contribution of REMs in NPPs for improving the performance and safety of the plants. Also, applications of various REMs are listed in Table 1.

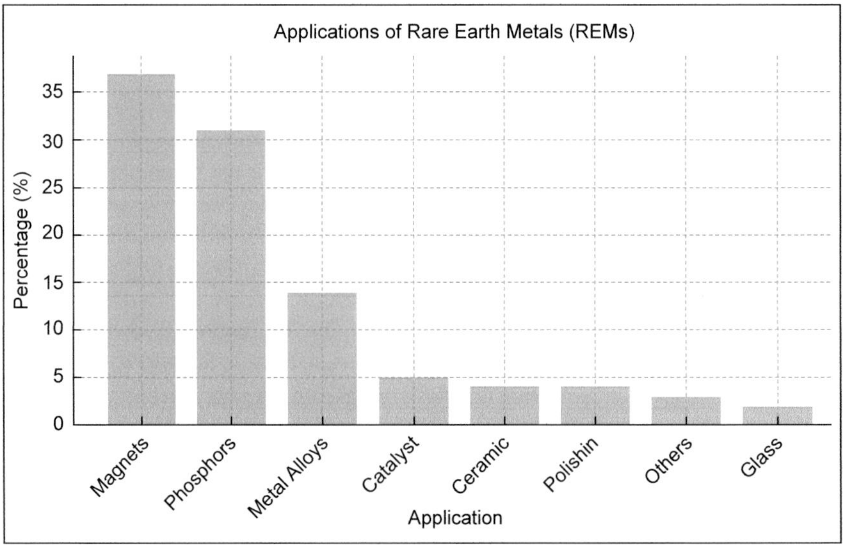

Fig. 2 Contribution of REMs in NPPs in terms of percentage.

Table 1 Applications of REMs in nuclear reactors.

Application	*Role of Rare Earth Elements*	*Reactor Type*
Fuel Components	Improve melting points and mechanical properties of actinide fuels	Fast Reactors, Sodium-Cooled Reactors (Zhao et al., 2023; Guon et al., 1990; Hossain et al., 2022; Kim et al., 2015; Chernov et al., 2025; Hossain et al., 2022; Yvon et al., 2015; Qiao et al., 2018; Igunma et al., 2024; Guo, 2024).
Neutron Absorbers	Regulate neutron flux in control rods and blankets	Pressurized Water Reactors (PWRs), Fast Reactors (Kim et al., 2015; Qiao et al., 2018; Igunma et al., 2024; Guo, 2024; Watkins, 2015; Malerba et al., 2022; Balbaud et al., 2021; Niarchos, 2021; Lee et al., 2013; Marra, 2011; Dumazert et al., 2018).
Radiation Shielding	Absorb gamma and neutron radiation in shielding glasses	All Reactor Types (Igunma et al., 2024; Guo, 2024; Watkins, 2015; Malerba et al., 2022; Balbaud et al., 2021; Niarchos, 2021; Lee et al., 2013; Marra, 2011; Dumazert et al., 2018; Borst et al., 2020).
Structural Materials	Provide radiation resistance and high-temperature strength in cladding and other components	Generation IV Reactors, Molten Salt Reactors (MSRs) (Igunma et al., 2024; Guo, 2024; Watkins, 2015; Malerba et al., 2022; Balbaud et al., 2021; Niarchos, 2021; Lee et al., 2013; Marra, 2011; Dumazert et al., 2018; Borst et al., 2020; Yu et al., 2018; Rafiuddin et al., 2022).

3. Challenges and Limitations

There are numerous advantages to adding REMs in nuclear reactors. However, there are still various limitations and challenges to using REMs in nuclear reactors, such as material degradation, fabrication complexity, supply chain constraints, radiation-induced damage, etc. REMs have properties to change their phase at a particular temperature, which converts into liquid and vapour that degrade the wall of the nuclear core. Containment of the NPPs consists of REMs which affect the thermal performance and efficiency of the NPPs (Guon et al., 1990). Also, by adding REMs in the reactor fuel, their fabrication requires advanced fabrication technologies and machines, which can increase costs and complexity (Igunma et al., 2024). Furthermore, the REMs are not easily available in the markets. So, their extraction and processing can have environmental and economic challenges (Guon et al., 1990; Hossain et al., 2022; Kim et al., 2015; Chernov et al., 2025;

Hossain et al., 2022; Yvon et al., 2015; Qiao et al., 2018; Igunma et al., 2024; Guo, 2024; Watkins, 2015). Also, due to the limited availability of the REMs, they are highly reactive and produce neutron radiation, which causes changes in their microstructure and properties and affects the NPPs (Cavazos et al., 2024).

4. Future Scopes and Research Opportunities for Researchers

There are various advantages, limitations, and challenges of using REMs in nuclear reactors and NPPs. Still, there is a lot of scope for researchers to develop novel rare earth-based materials for nuclear reactors which have higher mechanical strength, thermal sustainability, chemical stability, etc. Some areas are given below where researchers can focus such as advanced manufacturing methods, radiation resistance shields, sustainable supply chains, and waste minimization. Various emerging technologies are used to improve the mechanical and thermal properties of the REMs such as additive manufacturing and powder metallurgy, etc. These innovative fabrication methods can be economical and reduce difficult mechanics of producing rare earth-containing materials. Similarly, radiation resistance shields that are preferably made of REMs, such as high-entropy alloys and ceramic composites, are needed to optimize their performance in GIF-IV reactors (Igunma et al., 2024). Further, it is very difficult to extract the huge amount of REMs from the earth due to limited availability of the REMs. So, we have to develop sustainable and resilient supply-chain REMs for long-term use in NPPs and nuclear reactors (Guon et al., 1990; Watkins, 2015). However, by developing novel REMs or using existing REMs, lots of radioactive materials and energies are wasted. So, the use of REMs for minimizing nuclear waste, such as their application in advanced fuel cycles and waste disposal technologies, is expected to continue (Zhao et al., 2023; Guo, 2024; Watkins, 2015; Malerba et al., 2022; Balbaud et al., 2021; Niarchos, 2021; Lee et al., 2013; Marra, 2011; Dumazert et al., 2018; Borst et al., 2020; Yu et al., 2018; Rafiuddin et al., 2022; Cavazos et al., 2024; Reddy, 2024).

5. Conclusion

In the present work, an extensive review was carried out to identify various types of REMs that were used in NPPs and their applications in several sectors, as well as different methods to improve the REMs' properties for use in higher-temperature applications. REMs are very important materials that are extracted from the earth and used in one of the advanced energy source sectors, nuclear power plants, to improve fuel performance, neutron absorption, radiation shielding, and structural integrity. Also, REMs are used to increase the thermal efficiency of the NPPs and reduce waste by combining REMs with ceramic materials. They increase the thermal stability of the alloy material that is used in the construction of the core that is capable of withstanding higher temperatures and preventing the deterioration of the wall of the core and further use in the development of next-generation reactors.

However, various challenges of REMs, such as material degradation, fabrication complexity, and supply-chain constraints, are addressed to fully realize their potential. Also, the future scope of REMs in nuclear reactors and other fields is also discussed.

References

Balbaud, F., Cabet, C., Cornet, S., Dai, Y., Gan, J., Mayoral, M. and Weisenburger, A. (2021). An NEA review on innovative structural materials solutions, including advanced manufacturing processes for nuclear applications based on technology readiness assessment. *Nuclear Materials and Energy,* 27, 101006.

Bessais, L. (2021). Structure and magnetic properties of intermetallic rare-earth-transition-metal compounds: A review. *Materials,* 15(1), 201.

Borst, A.M., Smith, M.P., Finch, A.A., Estrade, G., Villanova-de-Benavent, C., Nason, P. and Geraki, K. (2020). Adsorption of rare earth elements in regolith-hosted clay deposits. *Nature Communications,* 11, 4386.

Cavazos, S.J., Younus, S., Uresti-Mireles, J., Sooby, E.S. and Piro, M.H.A. (2024). Rare earth carbide (Nd-C and Ce-C) synthesis and characterization to inform phase equilibrium in advanced nuclear fuels. *Superalloys,* 14(12), 1322.

Chernov, I.O., Kushtym, A.V. and Malykhin, S.V. (2025). Synthesis of materials based on compounds of rare earth elements with titanium, hafnium and zirconium as promising neutron absorbers for nuclear reactors. In: *Problems of Atomic Science and Technology*. ResearchGate. DoI: 10:46813/2024-152-064.

Chidambaram, R. and Banerjee, S. (2003). *Materials Research: Current Scenario and Future Projections*. Allied Publishers.

Dumazert, J., Coulon, R., Lecomte, Q., Bertrand, G.H. and Hamel, M. (2018). Gadolinium for neutron detection in current nuclear instrumentation research: A review. *Nuclear Instruments and Methods in Physics Research Section A*, 882, 53–68.

Elyassi, B., Rajabbeigi, N., Khodadadi, A., Mohajerzadeh, S.S. and Sahimi, M. (2004). An yttria-doped ceria-based oxygen sensor with solid-state reference. *Sensors and Actuators B: Chemical,* 103(1–2), 178–183.

Guo, D., Liu, J., Shi, J., Xu, B., Shi, B., Zhou, F. and Wang, Y. (2024). Synthesis and characterization of single-phase X2-Lu2SiO5 nanostructured feedstocks for advanced plasma-sprayed environmental barrier coatings. *Surface and Coatings Technology,* 476, 130230.

Guon, J., Doychak, J. and Licina, G.J. (1990). Advanced materials for nuclear reactor core components. *Journal of Nuclear Materials,* 171(1–2), 48–56.

Gupta, C.K. and Krishnamurthy, N. (1992). Extractive metallurgy of rare earths. *International Materials Reviews,* 37(1), 197–248.

Henderson, P. (Ed.). (2013). *Rare Earth Element Geochemistry*. Elsevier.

Hossain, M.K., Raihan, G.A., Akbar, M.A., Kabir Rubel, M.H., Ahmed, M.H., Khan, M.I. and El-Denglawey, A. (2022). Current applications and future potential of rare earth oxides in sustainable nuclear, radiation, and energy devices: A review. *ACS Applied Electronic Materials,* 4(7), 3327–3353.

Hossain, M.S., Rahman, M.M. and Kabir, M.A. (2022). Application of rare earth elements in nuclear reactors: A review of their roles in fuel, control, and structural materials. *Progress in Nuclear Energy,* 145, 104080.

Idemitsu, K., Arima, T., Inagaki, Y., Torikai, S. and Pouchon, M.A. (2003). Manufacturing of zirconia microspheres doped with erbia, yttria, and ceria by internal gelation process as a part of a cermet fuel. *Journal of Nuclear Materials*, 319, 31–36.

Igunma, T.O., Aderamo, A.T. and Chukwuemeka, O.H. (2024). Advanced corrosion-resistant materials for enhanced nuclear fuel performance: A conceptual review of innovations in fuel cladding against molten salt degradation. *Open Access Research Journal of Engineering and Technology,* 7(2), 016–030.

Kim, J.H., Kim, H.S. and Park, J.Y. (2015). Development of advanced nuclear materials for next-generation reactors. *Journal of the Korean Physical Society,* 67(4), 620–625.

Lee, W.E., Gilbert, M., Murphy, S.T. and Grimes, R.W. (2013). Opportunities for advanced ceramics and composites in the nuclear sector. *Journal of the American Ceramic Society,* 96(7), 2005–2030.

Malerba, L., Al Mazouzi, A., Bertolus, M., Cologna, M., Efsing, P., Jianu, A. and Tarantino, M. (2022). Materials for sustainable nuclear energy: A European strategic research and innovation agenda for all reactor generations. *Energies,* 15(5), 1845.

Marra, J. (2011). Advanced ceramic materials for next-generation nuclear applications. *Journal of the American Ceramic Society,* 18(16), 162001.

Niarchos, D. (2021). *Rare-earth high entropy alloys and transition metal high entropy alloys as building blocks for the synthesis of new magnetic phases for permanent magnet*s. Patent.

Patel, D.K., Sharma V.K. and Sharma, H.O. (2024). Advancements in aluminum matrix composites: A comprehensive review on the mechanical and microstructural attributes of rare earth elements (REEs) and rare earth oxides (REOs) doped materials. *Transactions of the Indian Institute of Metals*, 77(9), 2265–2276.

Patel, D.K., Sharma V.K. and Sharma, H.O. (2025). Current global Status of rare earth elements (REEs) and their role as catalysts in reducing air pollution for maintaining environmental sustainability. *Transactions of the Indian Institute of Metals*, 78(1), 4.

Patel, D.K. and Sharma V.K. (2025). Illuminating Innovations: A Comprehensive Review of Rare Earth-Doped Phosphors and Their Applications in LEDs. *Transactions of the Indian Institute of Metals*, 78(7), 165.

Qiao, J., Bao, M.L., Zhao, Y. J., Yang, H., Wu, Y., Zhang, Y., Hawk, J.A. and Gao, M.C. (2018). Rare-earth high entropy alloys with hexagonal close-packed structure. *Journal of Applied Physics,* 124(19), 195101.

Rafiuddin, M.R., Donato, G., McCaugherty, S., Mesbah, A. and Grosvenor, A.P. (2022). Review of rare-earth phosphate materials for nuclear waste sequestration applications. *ACS Omega,* 7(44), 39482–39490.

Rai, S.K., Kumar, P. and Panwar, V. (2021). Mathematical and numerical investigation of Ledinegg flow excursion and dynamic instability of natural circulation loop at supercritical condition. *Annals of Nuclear Energy,* 155, 108129.

Rai, S.K., Kumar, P. and Panwar, V. (2020). Numerical analysis of influence of geometry and operating parameters on Ledinegg and dynamic instability on supercritical water natural circulation loop. *Nuclear Engineering and Design,* 369, 110830.

Rai, S.K., Kumar, P. and Panwar, V. (2021). Numerical investigation of steady state characteristics and stability of supercritical water natural circulation loop of a heater and cooler arrangements. *Nuclear Engineering and Technology,* 53(11), 3597–3611.

Rai, S.K., Dutta, G. and Sheorey, T. (2017). Stability analysis of supercritical water natural circulation loop with vertical heater and cooler. In: *Proceedings of the 24th National and 2nd International ISHMT-ASTFE Heat and Mass Transfer Conference* (*IHMTC-2017*). Begel House Inc.

Rai, S.K., Kumar, P., Tiwari, M., Panwar, V., Kumar, D. and Sharma, V.K. (2025). A comprehensive overview of advancements, applications, and impact of supercritical fluid natural circulation loops. *Annals of Nuclear Energy,* 211, 110971.

Rai, S.K., Ahlawat, N., Upadhyay, R., Kumar, P. and Panwar, V. (2022). A study on the effect of geometry and operating variables on density wave oscillation in a supercritical natural circulation loop. *Computation,* 10(2), 25.

Rai, S.K., Kumar, P. and Panwar, V. (2020, November). Computational analysis of static flow instabilities in supercritical natural circulation loop. In: *Proceedings of International Conference on Thermofluids: KIIT Thermo 2020*, pp. 387–395. Springer Singapore.

Rai, S.K. and Dutta, G. (2018). A review of recent applications of supercritical fluid in natural circulation loops for nuclear reactor. *International Journal of Applied Engineering Research,* 23, 195–204.

Reddy, R.R. (2024). Prospective potential applications and emerging tendencies in rare earth materials. In: *Rare Earth: A Tribute to the Late Mr. Rare Earth, Professor Karl Gschneidner* (Vol. 164, pp. 177–210).

Sharma, V.K., Kumar, V. and Joshi, R.S. (2020). Parametric study of aluminium-rare earth based composites with improved hydrophobicity using response surface method. *Journal of Materials Research and Technology*, 9(3), 4919–4932.

Sharma, V.K., Kumar, V., Joshi, R.S. and Sharma D. (2020). Experimental analysis and characterization of SiC and RE oxides reinforced Al-6063 alloy based hybrid composites. *The International Journal of Advanced Manufacturing Technology*, 108(4), 1173–1187.

Sharma, V.K., Kumar, V., Joshi, R.S. and Kumar A. (2022). Effect of REOs on tribological behavior of aluminum hybrid composites using ANN. In: *Additive Manufacturing in Industry 4.0*, pp. 153–167. CRC Press.

Srikant, Sharma, V.K., Kumar, V., Joshi, R.S., Jain, S. and Kumar, P. (2021). A review of recent research on rare earth particulate composite materials and structures with their applications. *Transactions of the Indian Institute of Metals*, 74(11), 2569–2581.

Upadhyay, R., Rai, S.K. and Dutta, G. (2018). Numerical analysis of density wave instability and heat transfer deterioration in a supercritical water reactor. *Journal of Mechanical Science and Technology, 32*, 1063–1070.

Watkins, B.G. (2015). *Rare Earth Materials as Coating Compositions for Conductors* [Patent].

Yu, X. (2016). A kind of rare-earth-based nuclear radiation protection composite material and its preparation method (Chinese Patent No. CN105482225A). Shanghai Normal University. Published April 13, 2016.

Yvon, P., Le Flem, M., Cabet, C. and Seran, J.L. (2015). Structural materials for next generation nuclear systems: Challenges and the path forward. *Nuclear Engineering and Design,* 294, 161–169.

Zhao, S., Wang, P., Chen, W., Wang, L., Wang, Q.C. and Chen, W.Q. (2023). Supply and demand conflicts of critical heavy rare earth element: Lessons from gadolinium. *Resources, Conservation and Recycling,* 199, 107254.

9 Emerging Technologies in Rare Earth Element (REE) Recycling: A Path to Sustainable Resource Management

Soniya Goyal[1], Pardeep Kumar[2], Dinesh Kumar[3], Mahiti Gupta[4*], Vipin Kumar Sharma[5] and Mohd. Arezki Mellal[6]

1. Introduction

Rare earth elements (REEs) are one of the most needed elements in the 21st century that are found rarely in the earth's crust (Patel et al., 2025). It consists of 17 elements that are naturally available in the atmosphere (Srikant et al., 2021; Sharma et al., 2022). This group includes 15 lanthanides with two members from the scandium family also known as group I transition metals, viz., scandium, and yttrium (Patel et al., 2025; Patel et al., 2025) These elements are soft, malleable, shining remarkably like silver and when exposed to air they get oxidized slowly not abruptly (Patil et al., 2022). These rare elements and their alloys are important part of high-tech devices like compact fluorescent lamps, powerful magnets, wind turbines, smartphones, hard disk drives, hybrid and electric vehicles, and advanced defense technologies (Gaub et al., 2021; Sharma et al., 2019; Sharma et al., 2021). The widespread utilization of REEs is majorly due to their high atomic number and incompletely filled f-orbitals, which leads to quick transitions of electrons across the wide spectral range like from ultraviolet to infrared. These metals are in high demand due their distinctive catalytic, electromagnetic, chemical, and optical properties. These properties make these elements a requisite item for major high-tech devices needed for daily applications. Traditionally, REEs have been used in industries such as metallurgy, petrochemicals, textiles, and agriculture (Rybak & Rybak, 2022). REEs are widely used in permanent magnets containing around 30% of these elements. Permanent magnet is the top most application of these elements contributing around 37% of the total market (Nielen et al., 2023).

[1,4] Department of Biosciences and Technology, Maharishi Markandeshwar (Deemed to be) University, Mullana, Ambala.
[2,3] Department of Mechanical Engineering, Maharishi Markandeshwar (Deemed to be) University, Mullana, Ambala.
[5] Department of Mechanical Engineering, Meerut Institute of Engineering and Technology, Meerut, Uttar Pradesh, India.
[6] LMSS, Department of Mechanical Engineering, Faculty of Technology, M'Hamed Bougara University, Boumerdes, Algeria.
* Corresponding author: mahitigupta@gmail.com

DOI: 10.1201/9781003683254-9

1.1 Rare Elements Mining: A Difficult Task

Primarily to meet the needs of the present era is a big problem for REE mining. Around 2 lacs 80 thousand metric tons of REEs were mined in 2021, which was insufficient to meet the needs. Researchers have predicted that we will need around seven times more metal to meet the needs by the end of year 2040. The majority of demand is being fulfilled by China by producing around 2 lacs metric tons of rare earth oxides (Chen et al., 2024). Another difficulty is "so called balance problem" where the production of two REE metals, i.e., La and Ce are produced excessively but are not used so extensively whereas primary production of Nd and Dy are not able to meet the demands of their usage in magnets and batteries. This difficulty can be overcome by recycling the products containing Nd and Dy (Binnemans & Jones, 2015). Moreover, there are no substitutes for the REEs as the work done by them can't be contributed by any other element. Lastly, these elements do have rich deposits and are generally available as ores in earth's crust. So, miners have to use harsh treatments to separate them which leads to loss of element and increase in cost (Tian et al., 2025).

Although they are relatively abundant in the Earth's crust, their separation and extraction from ores pose significant scientific, technical, and environmental challenges. These difficulties arise primarily from the unique geochemistry of REEs, their complex mineral associations, and the nature of the chemical processes required for their separation (Rasool et al., 2025).

Geochemical Similarity of REEs: One of the primary reasons REE separation is difficult is their very similar chemical and physical properties. As members of the lanthanide series, REEs have nearly identical ionic radii and typically exist in the +3 oxidation state (Dong et al., 2021). This leads to nearly identical behavior in chemical reactions, making it extremely challenging to selectively extract one REE from another. Lanthanide contraction refers to the gradual decrease in ionic radii with increasing atomic number, but the differences are minimal (typically ~0.02 Å), making traditional separation techniques like precipitation or solvent extraction inefficient without significant processing (Cotruvo Jr., 2019). For example, neodymium (Nd^{3+}) and praseodymium (Pr^{3+}) are notoriously difficult to separate due to their closely matched ionic sizes and charge densities (Kumar et al., 2025).

Complex Mineralogy of REE-Bearing Ores REEs are not typically found in concentrated, easily processed forms. Instead, they occur in a wide variety of mineral phases such as bastnäsite (fluorocarbonates), monazite (phosphates), xenotime (yttrium phosphate), and others (Kumar et al., 2024). Each of these minerals has different chemical properties, which require distinct treatment routes. Monazite contains radioactive thorium, complicating its handling due to radiation safety and disposal requirements (Fabre et al., 2024). Bastnasite must be treated with strong acids or roasted at high temperatures to make REEs extractable, increasing energy consumption and environmental burden. In many deposits, REEs are dispersed across multiple mineral phases, requiring multi-stage physical and chemical treatments (Thomas et al., 2024). The application of finite element analysis (FEA) was used to evaluate the mechanical performance of GNP and CeO_2-reinforced AMCs for gear applications, demonstrating the relevance of simulation tools in virtual product development. Their work aligns with digital transformation in manufacturing, where computational modeling plays a key role in reducing material waste and optimizing component design (Kumar et al., 2025). Expanding the discussion, Kumar et al. (2024) presented a comparative review of AMC fabrication processes, including friction stir processing (FSP) and powder metallurgy, highlighting the role of advanced and sustainable processing technologies in tailoring composite microstructures. Their findings support the notion that manufacturing innovation directly impacts the functional versatility of composites in emerging applications.Focusing on surface engineering, Kumar et al. (2023) explored the friction stir processing of aluminum surfaces reinforced with GNPs, observing improved microstructural refinement and mechanical properties. This eco-friendly, solid-state processing technique illustrates the shift toward sustainable and energy-efficient fabrication methods.In terms of corrosion resistance, Kumar et al. (2023) evaluated stir-cast Al6061-CeO_2 nanocomposites in NaCl and H_2SO_4 environments, reporting enhanced durability due to the barrier effect of CeO_2. Furthermore, Kumar et al. (2023) investigated hybrid reinforcement (GNPs + CeO_2), which provided both corrosion resistance and structural reinforcement, showing promise for marine and chemical industry applications.The mechanical and wear performance of AMCs was reviewed by Kumar et al. (2022), who emphasized the impact of reinforcement morphology, distribution, and interfacial bonding on wear resistance and structural integrity. Their review reinforces the trend of material optimization through microstructural control, an essential aspect of smart materials. On the thermal front, Kumar et al. (2025) developed a probabilistic thermodynamic model to predict valve temperature transients in diesel engines. Although not strictly within AMC fabrication, this study contributes to emerging thermal management technologies, relevant for components made with thermally conductive composites. Complementary to these material innovations, research in supercritical natural circulation loops (SCNCLs) has expanded significantly. Rai et al. (2022) examined how geometry and operating conditions affect density wave oscillations in SCNCLs using computational simulations, providing insights into the instability behavior of supercritical fluids—a key concern for the design of safe and efficient nuclear thermal systems. Building on this foundation, Rai et al. (2025) presented a comprehensive review of SCNCL advancements, applications, and their potential impact on next-generation reactors. The review highlights SCNCLs as a passive and efficient thermal transport technology, essential for future energy systems that demand both compact design and high heat flux capabilities without relying on active pumping.

Refractory Nature and Low Concentration: The concentration of REEs in ores is typically low, often less than 10% by weight, and their strong association with gangue materials makes them refractory (resistant to dissolution). This demands extensive beneficiation steps (crushing, grinding, flotation, magnetic separation), followed by hydrometallurgical processing involving harsh chemicals (e.g., concentrated HCl, H_2SO_4, NaOH) and multiple purification cycles (Abadi & Musci, 2024). The chemical resistance of minerals like xenotime or allanite necessitates high-temperature digestion or fusion. These processing steps generate large volumes of toxic waste, including acid tailings and radioactive residues (Balaram, 2023).

Environmental and Regulatory Challenges: The environmental footprint of REE extraction is significant due to the consumption of acids, alkalis, water, and energy, along with the generation of radioactive and hazardous wastes. This makes the permitting and operation of REE extraction facilities both economically and politically challenging. China's dominance in REE production has been linked to lax environmental regulations, which are difficult to replicate elsewhere due to higher regulatory standards (Patel et al., 2024). Attempts to open REE mines in the U.S., Canada, and Australia have often faced strong opposition due to concerns about environmental degradation and radioactive by-products.

Separation Technologies are Expensive and Time-Consuming: To separate REEs at high purity levels (> 99.9%), industrial processes rely on solvent extraction and ion-exchange chromatography, which involve dozens to hundreds of stages due to the REEs' chemical similarity. Solvent extraction, the most widely used method, uses complexing agents and organic solvents to achieve phase transfer between aqueous and organic layers (Kurkinen et al., 2023). The process is highly selective but requires multiple stages and precise control of pH, temperature, and flow rates. Ion-exchange methods are used more for analytical or laboratory-scale purification due to high cost and slower processing times (Kumar et al., 2023).

The difficulty in separating REEs from ores is rooted in their geochemical similarity, complex mineralogy, low concentration, and the environmental impact of extraction processes. Advanced separation methods, though effective, are capital- and labor-intensive, requiring sophisticated infrastructure and regulatory compliance. These challenges collectively explain the limited number of economically viable REE production facilities worldwide and highlight the urgent need for cleaner and more efficient extraction technologies (Gikka et al., 2024).

2. Recycling: The Game Changer

Recycling would meet up to 1/4th of the need for REEs in about 10 years. Recycling REEs from end-of-life (EoL) products offers a promising strategy to reduce environmental impact, enhance supply security, and promote circular economy practices. Key sources for recycling include permanent magnets (e.g., from hard drives and wind turbines), nickel-metal hydride batteries, fluorescent lamps, and electronic waste. However, the recycling of REEs presents several challenges due to the diversity of product compositions, the presence of complex material matrices, and the difficulty in separating individual rare earths from one another

(Dagwar et al., 2025). To address these challenges, various recycling technologies are being developed and optimized, including hydrometallurgical processes (acid leaching and solvent extraction), pyrometallurgical methods (smelting and roasting), bioleaching, and emerging green technologies like ionic liquids and supercritical fluids. These methods aim to recover REEs efficiently, selectively, and with minimal environmental burden. As the transition to clean energy and digital technologies accelerates, the development of scalable, economically viable, and environmentally friendly recycling processes will be crucial (Hamzat et al., 2025). Integrating REE recycling into national resource strategies and industrial practices is essential for building a resilient and sustainable supply chain for the future.

2.1 Importance of Recycling REEs

The strategic importance of rare earth elements cannot be overstated. As key enablers of clean energy and advanced technologies, REEs are critical to the development of wind turbines, electric vehicles, high-efficiency motors, and electronics. The growing demand for these materials—driven by global efforts toward decarbonization and digitalization—has raised concerns about the sustainability and ethical implications of conventional mining practices.

Recycling provides an essential pathway to reduce environmental impacts, lower dependency on primary sources, and diversify supply chains. It also allows the recirculation of high-purity REEs back into manufacturing without the need for extensive beneficiation or processing of raw ores.

2.2 Major Sources of Recyclable REEs

REEs are present in a wide range of consumer and industrial products. The most significant sources for recycling include:

- **NdFeB Magnets:** Used in hard drives, electric motors, wind turbines, and headphones. These magnets contain neodymium (Nd), praseodymium (Pr), dysprosium (Dy), and terbium (Tb) (Xiao et al., 2023).
- **Nickel-Metal Hydride (NiMH) Batteries:** Contain lanthanum (La) and cerium (Ce), and are used in hybrid vehicles and electronics.
- **Fluorescent Lamps and CRTs:** Contain phosphors doped with europium (Eu), yttrium (Y), terbium (Tb), and other REEs.
- **Catalytic Converters and Polishing Powders:** Used in automotive and optical industries.

2.3 Traditional Recycling Techniques of Rare Earth Elements (REEs)

Traditional recycling of REEs refers to the well-established, industrially practiced methods developed over past decades to extract and recover REEs from EoL products or production waste. These techniques mainly include hydrometallurgical and pyrometallurgical processes, which have long been used for primary ore processing and later adapted to secondary (recycled) sources (Cherkezova-Zheleva et al., 2024). Although not always environmentally benign or economically optimal, these processes have provided a basis for recovering valuable REEs from scrap materials, particularly NdFeB magnets, phosphors, and nickel-metal hydride batteries.

2.3.1 Hydrometallurgical Techniques

Hydrometallurgy is the most widely used traditional method for the recovery of REEs. It relies on the use of aqueous chemistry to dissolve, separate, and purify REEs from complex waste matrices. The hydrometallurgical recycling of REEs typically involves the following sequential steps (Zhang et al., 2020) represented in Figure 1. Each of these steps is highly dependent on the input material's composition and the desired purity of the final REE product.

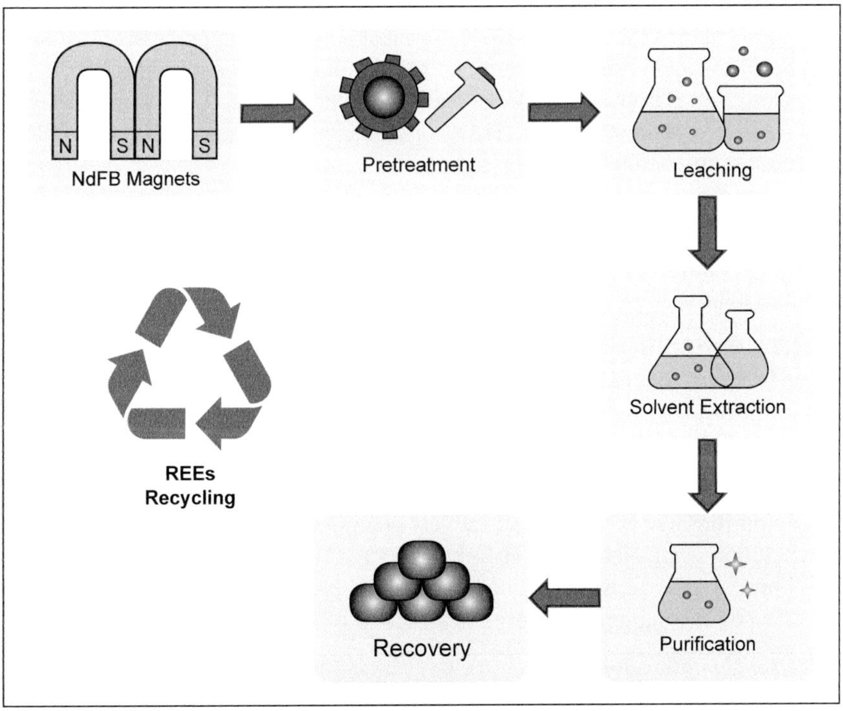

Fig. 1 Pictorial representation of steps used in Hydrometallurgy.

2.3.1.1 Leaching

It is the initial and critical step where REEs are brought into solution from solid matrices. It involves contacting the material with an appropriate leachant to dissolve REE ions while minimizing co-dissolution of impurities.

Common Leaching Agents

- Inorganic Acids:
  - Hydrochloric acid (HCl): Efficient for dissolving REEs from NdFeB magnets and phosphors. It enables high REE solubility and fast reaction kinetics.
  - Sulfuric acid (H_2SO_4): Widely used due to low cost; common in lamp phosphor recycling.
  - Nitric acid (HNO_3): Strong oxidizer, used for phosphors but less common due to expense and fumes.
- Alkaline Leaching:
  - Applied particularly for materials like red lamp phosphors ($Y_2O_3:Eu^{3+}$) or cerium-rich ores, where REEs are resistant to acid dissolution.
  - NaOH or Na_2CO_3 may be used to selectively dissolve certain rare earth compounds.

Leaching Parameters

- Temperature: Typically 50–90°C for most acid leaching processes.
- Solid-to-liquid ratio: Determines concentration of REEs in solution.
- Time: Varies from minutes to several hours depending on material and leachant.
- Additives: Reductants (e.g., H_2O_2) or oxidants may be added to enhance REE dissolution or alter oxidation states.

In this step, REEs are solubilized from solid waste using strong acids or alkalines. For example, NdFeB magnets are crushed and treated with HCl, resulting in the dissolution of neodymium and praseodymium, leaving behind iron-rich residues (Yun et. al., 2024).

2.3.1.2 Purification

After leaching, the resulting solution often contains impurities such as Fe, Al, Zn, or transition metals. These must be removed before REEs can be effectively separated (Soltani et al., 2021).

pH Adjustment and Precipitation

- Impurities like Fe^{3+} can be selectively precipitated by increasing the pH of the solution using NaOH or NH_4OH.
- For example, iron precipitates as $Fe(OH)_3$ around pH 3.5–4.5, while REEs remain in solution.

Selective Complexation

- Certain ligands or complexing agents may be added to form stable complexes with non-REE metals, enabling easier separation.

2.3.1.3 Solvent Extraction

Due to the chemical similarity of REEs, particularly their trivalent oxidation state, separation into individual elements is one of the most challenging and critical steps (Yun et. al., 2024).

A classical and highly effective method using organic solvents to separate REEs based on subtle differences in ionic radius and charge density.

- Extractants:
  - Acidic Organophosphorus Compounds:
    - D2EHPA (di-(2-ethylhexyl) phosphoric acid)
    - PC88A and Cyanex 272 are commonly used for separating light from heavy REEs.
  - Neutral or Basic Extractants:
    - Aliquat 336, TBP (tri-butyl phosphate) for certain REEs or actinides.
- Process:
  - Aqueous REE solution is contacted with organic extractant.
  - REEs transfer to the organic phase; impurities stay in the aqueous phase.
  - Multiple counter-current stages may be employed to improve separation efficiency.
- Ion Exchange
  - Ion exchange resins can selectively adsorb REEs based on ionic size and affinity.
  - Suitable for high-purity separation in small-scale applications.
  - Resins are regenerated by elution with acid or salt solutions.

2.3.1.4 Precipitation and Calcination

Once the REEs are separated and concentrated, they are recovered as solid compounds, usually by precipitation (Yun et. al., 2024).

Oxalate Precipitation

- **Oxalic acid ($H_2C_2O_4$)** is added to the REE-rich solution to form insoluble REE oxalates:

$$REE^{3+} + 3C_2O_4^{2-} \rightarrow REE_2(C_2O_4)_3\downarrow$$

Carbonate or Hydroxide Precipitation

- **Sodium carbonate (Na_2CO_3)** or **ammonium hydroxide (NH_4OH)** may also be used for REE recovery depending on downstream processing needs.

Calcination

- Precipitated REE salts are calcined at ~800–1000°C to convert them into REE oxides (REOs), which are marketable forms for reuse.

2.3.2 Pyrometallurgical Techniques

Pyrometallurgical recycling involves the treatment of REEs containing materials at elevated temperatures to recover valuable metals. These techniques are well-established in primary metallurgy and have been adapted for recycling due to their ability to process large volumes of waste, especially from industrial sectors like NdFeB magnet manufacturing, spent catalysts, and electronic waste (Holzer et al., 2022). Unlike hydrometallurgy, which relies on aqueous chemistry, pyrometallurgy utilizes thermal processes such as smelting, roasting, and distillation to liberate and separate REEs from complex matrices.

2.3.2.1 Direct Smelting and Slagging

Principle: Smelting involves melting the feed material at high temperatures (1300–1600°C), during which REEs typically form stable oxides or alloy phases. Iron and other non-REE metals can be separated as a metallic or slag phase.

Application:

- Used for magnet scrap and production waste (e.g., NdFeB alloys).
- Iron is oxidized and removed as slag; REEs concentrate in the metallic phase or the slag, depending on the flux and atmosphere used.

Process Steps:

- Crushing and pre-treatment of magnet scrap or residues.
- Addition of fluxing agents (e.g., CaF_2, SiO_2) to promote slag formation.
- Smelting in an induction furnace or arc furnace.
- Separation of metal and slag phases.
- Further treatment of slag or metal to extract individual REEs via hydrometallurgy.

Advantages:

- Handles large and impure materials
- Rapid processing with minimal chemical reagents

Limitations:

- High energy consumption
- Loss of REEs into the slag phase unless carefully optimized
- Not effective for all REE-containing waste streams (e.g., phosphors)

2.3.2.2 Selective Oxidation and Reduction

Principle: Oxidizing or reducing environments are used to convert specific elements in the waste into different oxidation states or compounds, thereby allowing separation.

- Selective oxidation: Iron is preferentially oxidized and removed, leaving REEs in metallic or stable oxide form.
- Selective reduction: REEs can be reduced to metallic form, separating them from oxidized impurities.

2.3.2.3. Chlorination and Halide Volatilization

Principle: Some REEs form volatile chlorides at elevated temperatures. By converting REEs to gaseous chlorides (e.g., $NdCl_3$, $DyCl_3$), they can be selectively volatilized and recovered by condensation.

Process:

1. Heating REE-bearing material in the presence of Cl_2 or NH_4Cl at 700–1000°C.
2. Formation and volatilization of REE chlorides.
3. Condensation of vapors to recover REEs.

Advantages:

- High selectivity for some REEs (e.g., Dy, Nd).
- Effective separation from less volatile impurities.

Challenges:

- Handling of corrosive chlorine gas or ammonium chloride.
- Limited industrial implementation due to safety concerns and gas recovery requirements.

2.3.2.4 Vacuum Distillation and Sublimation

Principle: Certain REEs (especially heavy REEs like Dy and Tb) exhibit differences in vapor pressure at high temperatures. These can be exploited under vacuum to selectively distill them from alloys or magnet scrap.

Typical Application:

- NdFeB magnets containing Dy are treated at 1200–1400 °C under vacuum ($\sim 10^{-3}$ Pa).
- Dy sublimes and is condensed onto a cold surface, separating it from other REEs and Fe.

Advantages:

- No chemicals required—physical separation only.
- Can achieve high-purity separation of specific REEs.

Limitations:

- High equipment cost and energy consumption.
- Not suitable for low-value or dilute waste streams.

2.3.3 *Limitations of Traditional Techniques*

Often these REE have been blended with other metals. Separating them out again can be very hard. In some ways, recycling rare earths from tossed-out items is as challenging as extracting them from ore and processing them.

Rare-earth recycling tends to use hazardous chemicals, such as hydrochloric acid. It also uses a lot of heat, and thus a lot of energy. And that effort may only recover a tiny amount of metal (Gkika et al., 2024).

While traditional recycling processes have shown technical feasibility, several limitations hinder their widespread adoption:

- *Environmental Concerns:* Use of concentrated acids and organic solvents generates toxic waste and emissions.
- High Operating Costs: Complex separation steps and chemical consumption increase processing expenses.
- Low REE Yield from Mixed Waste: Efficiency drops significantly when dealing with heterogeneous or miniaturized electronics.
- Lack of Standardization: Dismantling and pre-treatment processes vary depending on product design and manufacturer.

3. Outlook and the Need for Innovation

Traditional REE recycling techniques have laid the groundwork for secondary resource recovery but must evolve to meet modern environmental and economic goals. Recent efforts focus on enhancing process selectivity, reducing waste, and incorporating green chemistry principles. Hybrid approaches—combining pyro- and hydrometallurgy or integrating physical and chemical separation—are being explored. There is also increasing emphasis on designing REE-containing products for easier EoL disassembly and material recovery, a concept known as design for recycling (DfR).

Going forward, traditional recycling techniques will likely serve as the backbone of REE recycling infrastructure, while being supplemented by emerging technologies such as bioleaching, ionic liquids, and electrochemical methods. Scaling these systems efficiently and sustainably is critical to building a secure, circular REE supply chain.

4. Emerging and Green Technologies

4.1 Ionic Liquid and Deep Eutectic Solvents

The development of environmentally friendly and sustainable alternatives for the extraction and recycling of REEs has become mandatory due to the increasing demand for these elements worldwide. The conventional methods such as solvent extraction, pyrometallurgy, and hydrometallurgy are efficient but they frequently employ toxic chemicals, emit hazardous waste, and demand large amounts of energy. Green solvent systems include ionic liquids (ILs) and deep eutectic solvents (DESs) have become viable alternatives for REE separation and purification in response to these limitations due to their adjustable characteristics, improved selectivity, and reduced volatility (Abbott et al., 2003; Binnemans et al., 2013). A safer and more sustainable approach of extracting metals is offered by ionic liquids, which are non-volatile, thermally stable, and recyclable solvents composed completely of ions. To selectively recover specific REEs from complicated matrices such as electronic trash, discarded catalysts or manufacturing slags, they can be tailored at the molecular level. Under mild conditions, ILs has proven to be quite effective at leaching and extracting REEs, which minimizes consumption of energy and impact on the environment (Yin et al., 2013). Furthermore, they improve process efficiency by dissolving both organic and inorganic compounds. To optimize REE selectivity, a variety of task-specific ILs have been designed, their application in supported liquid membranes and hybrid solvent extraction systems further enhances the efficacy of metal recovery (Binnemans et al., 2013).

A subclass of IL analogues known as deep eutectic solvents, which are usually developed by addition of hydrogen bond donor and acceptor, offer similar benefits along with additional advantages of being less expensive, simpler to produce and biodegradable. DESs are extremely desirable due to their non-toxicity, sustainability, and simplicity of manufacture from bio-based or renewable components, such as organic acids and choline chloride (Smith et al., 2014). Choline chloride-lactic acid or choline chloride-urea eutectics are extensively studied examples that have shown effective dissolving of NdFeB magnets, a remarkable secondary source of neodymium and dysprosium (Van den Bosch et al., 2019). These eutectics help in the selective leaching of rare earths by minimizing the solubility of iron and other base metals. Without the requirement of harsh acids or high temperatures, they effectively leach from ores and waste streams because of their beneficial complexation abilities with REEs. The sustainability of REE recovery can be enhanced by integrating DES-based systems with other green technologies and exhibiting their compatibility with bioleaching methods (Zhang et al., 2022). By diminishing the utilization of hazardous materials and facilitating the reuse and recycling of valuable elements, DES and ILs both act as an excellent example of the notions of green chemistry and the circular economy. Their development is in accordance with the strategy objective of minimising dependency on primary REE sources and diminishing environmental hazards related to conventional methods of extraction. In order to enhance the practical viability of these solvents, more research should be carried out regarding how to effectively optimize them for industrial scale purposes, enhance their recyclability, and minimize production costs. Thus, in the aim of sustainable REE recycling, DESs and ILs are distinguished as breakthrough technologies.

4.2 Bioleaching

Microbial leaching, also known as bioleaching, is becoming ever more popular as an eco-friendly and sustainable approach of recycling REEs. In this technique, microorganisms such as bacteria and fungi are used for extraction of metals from solid matrices including low-grade ores, mining tailings, industrial waste, and EoL products such as magnets and electronic components. Bioleaching is an alternative approach in green technology in the framework of the circular economy since it does not need any hazardous chemicals, significant amount of energy or a large infrastructure as compared to conventional methods like pyrometallurgy and hydrometallurgy (Brandl et al., 2018). In case of REEs, bioleaching utilizes the ability of specific microorganisms to produce organic acids or chelating agents which solubilize rare earth metals. *Aspergillus niger* and *Penicillium simplicissimum* produce citric, oxalic, and gluconic acids which are known to effectively leach lanthanides from phosphorus, spent catalysts, and NdFeB magnets (Ilyas et al., 2020). Acidophilic bacteria *Acidithiobacillus ferrooxidans* has also demonstrated the ability to mobilize REEs by oxidizing sulphide minerals and promoting the production of related rare earth ions. These biological systems stimulate selectivity towards REEs while simultaneously minimizing the environmental impact of leaching via specific biochemical pathways. One of the noteworthy uses of bioleaching is the recovery of yttrium and europium from red phosphor powders and fluorescent bulbs which have reached the end of their useful lives. According to studies, pre-treated biomass or mixed microbial cultures can remarkably enhance the dissolving efficiency of these elements without the use of potent mineral acids (Park & Lee, 2021). Similarly employing fungi which synthesize citric acid to bioleach REEs from NdFeB magnets has shown significant outcomes, particularly when combined with mechanical pre-treatment methods that enhance surface area and bioavailability (Bosshard et al., 2016).

Although bioleaching has several advantages; its industrial-scale application is limited due to few limitations including relatively slow kinetics of metal solubilization, limited tolerance of microorganisms to heavy metal concentrations, and need to optimize growth conditions such as temperature, pH, and nutrient availability. Moreover, the specificity and efficiency of microbial action is frequently impacted by the composition of waste products. To overcome these challenges, current research is focusing on the development of synergistic systems which combine bioleaching with green solvents such as DESs and genetically modified microorganisms with enhanced metal tolerance and acid production potential (Zeng et al., 2022).

In conclusion, bioleaching is a crucial component of emerging green technologies for recycling rare earth elements. It is especially suitable for incorporating into sustainable resource management strategies because of its compatibility with decentralized and low-impact processing technologies. Bioleaching is projected to emerge as an effective approach for recovering REEs in a sustainable manner with further development in microbial biotechnology and bioprocess optimization.

4.3 Supercritical Fluid Extraction

In the recycling of REEs, supercritical fluid extraction (SFE) is an innovative and sustainable approach that is becoming immensely prominent. To accomplish enhanced solubility, diffusivity, and selectivity, this technique employs fluids at temperatures and pressures above their critical points, where they exhibit distinctive physicochemical characteristics situated between gases and liquids. Supercritical carbon dioxide ($scCO_2$) is the most studied among supercritical fluids due to its non-toxic, non-flammable, affordable, and easily accessible behavior. SFE is consistent with the concepts of green chemistry by minimizing the requirement for hazardous organic solvents, producing less waste and functioning under generally mild conditions (Pourret et al., 2018). SFE is commonly used together with complexing agents or ligands in REE recycling to enhance the extraction efficacy of lanthanides and other rare earth metals from solid matrices such as electronic waste, phosphors, and NdFeB magnets. While $scCO_2$ is non-polar and has limited solubility for ionic compounds like REEs, but when treated with suitable chelating agents, such as organophosphorous extractants, i.e., tributyl phosphate, its metal affinity significantly enhance (Subramanian et al., 2015). This approach provides a cleaner alternative for conventional solvent extraction methods by facilitating the selective separation of particular REEs. SFE is especially advantageous for processing waste with complicated matrices due to great separation efficacy, and renders solvent recovery and recycling simple.

The simplicity of post-extraction processing is a remarkable advantage of SFE. The supercritical fluid turns into a gas by simply decreasing the pressure, leaving behind REE extract in concentrated form and removing the requirement of solvent removal procedures. Furthermore, by varying the extraction conditions, the pressure and temperature adaptability enables the selective targeting of various REEs. SFE can be used for customized separation procedures in multi-element waste streams, such as used automotive catalysts or EoL electronics due to their manageability (Li et al., 2020). Despite its immense potential, SFE is still at the research and pilot-scale stage for REE recovery. Some of the limitations associated with $scCO_2$ include relatively lower extraction yield as compared to aqueous systems, expenses of equipment and the development of highly efficient ligands compatible with $scCO_2$. However, these challenges are being promptly addressed by advancements in hybrid SFE systems, designing of green ligands, and process intensification. By integrating SFE with other green technologies such as microwave-assisted leaching, deep eutectic solvents, or ionic liquids may open the door to integrated and sustainable REE recycling systems (Cheng et al., 2022).

In summary, SFE is a proactive approach to long-term REE recovery. It is an attractive option in the portfolio of green technologies intended to satisfy the rising demand for REEs globally due to its low environmental impact, solvent-free residue, and scalability potential. SFE may soon transition from laboratory innovation to industrial application in the field of resource recovery and key material sustainability due to continuous advancements in technology.

4.4 Electrochemical Methods

REEs can be recycled and extracted from a variety of waste and secondary resources using electrochemical methods, which are evolving as economically feasible and sustainable technologies. These approaches, which usually do not require harsh chemicals or high temperatures, employ the principles of electrochemistry to particularly reduce, oxidize, or deposit target metals under specified electrical conditions. Electrochemical methods are cleaner and energy efficient alternatives as compared to traditional metallurgical operations for treating complex waste matrices which include spent batteries, electronic debris, and EoL magnets (Zhang et al., 2021).

The capability of electrochemical methods to achieve excellent selectivity for REEs through potential control is a major advantage, allowing the targeted recovery of particular elements while eliminating the co-extraction of contaminants. Neodymium, dysprosium, and other lanthanides can be recovered from NdFeB magnets and phosphor wastes through electrochemical techniques such as electro deposition, electrowinning, and electrochemical leaching. For instance, direct metal recovery from molten salts or chloride-based aqueous electrolytes without the requirement of further precipitation or solvent extraction procedures is possible using electrodeposition (Kim et al., 2018). Furthermore, the efficacy and selectivity of these methods have been enhanced by the advancement of sophisticated electrode material and redox mediators, providing the way for their integration into closed-loop recycling systems.

Another innovative approach for improving the dissolution of REEs from matrices is electrochemical leaching, which employs electrical energy. As compared to chemical leaching, this method can function at room temperature and can be adjusted according to particular dissolving kinetics by varying the voltage and current parameters. Some research has shown that utilizing ionic liquids or deep eutectic solvents as electrolytes in electrochemical setups further improves the sustainability and selectivity of process by substituting corrosive mineral acids with green solutions (Tunsu et al., 2015). These hybrid technologies not only lessen negative impacts on the environment but also allow simultaneous leaching and separation of REEs from mixed metal waste. One of the main advantages of using electrochemical processes is their capacity to scale and adapt to modular, decentralized recycling units. Small-scale electrochemical reactors can be employed in local e-waste processing facilities or urban mining operations, aiding to decentralize and circularize supply chains for critical materials. To ensure economic and technical viability at industrial scales, some challenges such as electrode corrosion, optimization of energy consumption, and recovery of electrolyte must be addressed (Amaral et al., 2019).

In summary, electrochemical methods are a vital and adaptable part of emerging green technologies for recovering REEs. They are an effective choice for the sustainable recovery of REEs because of their compatibility with the environment, process flexibility, and capacity for integration with renewable energy sources. With ongoing developments in selectivity, efficacy, and economic viability of electrochemical systems, these technologies have the potential to contribute a significant part in stimulating sustainable resource management in the rare earth industry.

4.5 Membrane and Adsorption Technologies

Adsorption and membrane technologies are gaining more attention as effective and sustainable alternatives for recycling REEs. These technologies are suitable for decentralized and sustainable resource management, hence they offer selective recovery mechanisms without the utilization of hazardous chemicals. Membrane-based separation allows some ions or molecules to flow through semi-permeable barriers while hindering others by using physical and chemical selectivity. The applications of supported liquid membranes (SLMs), ultrafiltration, and nanofiltration have shown potential in the separation and concentration of REEs from leachates derived from industrial effluents, end-of-life magnets, and electronic waste (Pan et al., 2021). The developments in membrane materials include polymer inclusion membranes and nanocomposite membranes which leads to further enhancement in ion selectivity, permeability, and chemical stability.

The SLMs, among all membrane types, may combine solvent extraction and membrane separation into a single process; they are particularly suitable for the extraction of REEs. SLMs allow stringent selective transport of lanthanides across the membrane via immobilization of a carrier extractants like D2EHPA or

TOPO inside a porous membrane support (Zhao et al., 2020). Furthermore, the incorporation of deep eutectic solvents and ionic liquids into membrane systems leads to the improvement in the efficacy and environmental safety of these technologies by removing the need for volatile organic solvents.

Adsorption-based technology is another important class which includes scalable, low-energy methods for extraction and recovery of REEs. In these technologies, REEs are attached to solid sorbents which have huge surface area and some functional groups. A wide range of substances have shown significant affinity for REEs include metal-organic frameworks (MOFs), activated carbon, ion-exchange resins, biosorbents, and functionalized silica (Wang et al., 2021). Biosorbents based on alginate and phosphorylated cellulose have shown remarkable potential in the selective adsorption of light and heavy REEs from intricate aqueous matrices. Due to its reusability, affordability, and ease of handling, adsorption is a suitable method for industrial and post-consumer REE waste streams. Adsorption-based technologies have been further refined by recent developments in magnetic adsorbents and nano-engineered sorbents. An external magnetic field can be applied to rapidly separate magnetic nanoparticles functionalized with chelating ligands from solution, enhancing effective cycles of separation and recovery. Similarly, chemical functionalities and personalized pore architectures are offered by functionalized MOFs, and hybrid materials are used to extract specific REE ions from diluted and competing environments (Huang et al., 2023). Moreover, it has been studied that hybrid methods which integrate adsorption and membrane systems enhance output and selectively aid to intensify the REE recycling process.

Research is still ongoing to find alternatives through system optimization and new material development to address different challenges such as membrane fouling, adsorbent regeneration efficiency, and scalability associated with both membrane and adsorption systems. In summary, membrane and adsorption technologies help in transition to more sustainable resource management by offering adaptable, scalable, and sustainable methods for recovering vital REEs. Comparison of traditional recycling techniques and emerging technologies have been given in Table 1.

Table 1 Comparison of REE Recycling Technologies.

Technology	*Recovery Rate*	*Environmental Impact*	*Cost*	*Scalability*
Hydrometallurgy	80–98%	Moderate–High (acid use)	Moderate	High
Pyrometallurgy	60–85%	High (energy intensive)	High	Medium
Ionic Liquids	70–95%	Low	High (solvent cost)	Medium
Bioleaching	40–70%	Very Low	Low	Low–Medium
Membrane Separation	60–90%	Low	Moderate	Medium

4.6 Microwave-assisted Leaching

Microwave-assisted leaching is also used as a promising green method to enhance the effectiveness of recovering REEs from primary and secondary sources. As compared to conventional thermal leaching techniques, microwave heating directly impacts the material's dielectric features to produce rapid, volumetric heating across a solid matrix. Microwave-assisted leaching is a very prominent alternative for sustainable REE recycling; it leads to remarkable reduction in processing time, reagent consumption, and energy use (Zhu et al., 2021). The efficacy of microwave-assisted leaching lies in its potential to disintegrate the crystal lattice and increase reagent penetration, which facilitates metal dissolution. This technique is mainly beneficial in processing REE-rich waste including wasted fluorescent lamps, NdFeB magnets, red phosphor powders, and weathered ores. Studies have shown that microwave pre-treatment followed by acid or chelating agent leaching significantly stimulates the extraction yields of neodymium, europium, dysprosium, and yttrium under milder conditions as compared to conventional techniques (Abhilash et al., 2017).

Furthermore, microwave-assisted leaching is compatible with green solvents such as organic acids, ionic liquid, and deep eutectic solvents, and lessens reliance on hazardous mineral acids. According to the study, this amalgamation promotes the objectives of circular economy and green chemistry frameworks and reduces the amount of secondary pollutants (Yang et al., 2020). By modifying microwave parameters such as power density, frequency, and exposure period, a substantial amount of control over process kinetics and elemental separation is also possible which can be used to optimize the specificity of leaching processes. Despite its apparent benefits, the scalability of microwave-assisted leaching is still a limitation. Thus, it is mandatory to overcome limitations such as energy optimization, equipment designing, and uniform heating associated with technology for industrial purposes. Hybrid systems which combine microwave treatment with ultrasonic or electrochemical processes along with new advancements in microwave reactor engineering are paving the way to more practical commercial deployment (Chen et al., 2022).

In summary, microwave-assisted leaching is serving as an effective and sustainable method for recovering REEs. It is a crucial part of future-ready rare earth recycling schemes due to its rapid kinetics, low usage of reagents, and compatibility with green solvents. Microwave-assisted technology is projected to be a vital component of the transition to decentralized and sustainable REE resource management with additional research and process development.

5. Conclusion

The recycling of rare earth metals is essential for ensuring sustainable resource management, reducing dependency on primary mining, and mitigating environmental impacts. Traditional techniques such as hydrometallurgical and pyrometallurgical methods have been widely adopted due to their technical maturity and efficiency in recovering REEs from EoL products. However, these approaches often involve high energy consumption, use of hazardous chemicals, and generation of secondary waste.

In response to these challenges, emerging technologies such as bioleaching, supercritical fluid extraction, electrochemical methods, and ionic liquid-based systems are gaining attention. These methods offer significant advantages including selectivity, lower environmental impact, and potential for scalability. Additionally, advancements in sensor-based sorting, mechanical pretreatment, and automated disassembly are enhancing the efficiency of REE recovery from complex waste streams like NdFeB magnets, phosphors, and electronic waste.

While no single technique is universally superior, a hybrid approach combining traditional and emerging methods, tailored to specific waste sources and economic contexts, holds the greatest promise. Continued research, policy support, and industrial collaboration are crucial to transition from laboratory-scale innovations to commercially viable and environmentally responsible REE recycling solutions.

References

Abbadi, A. and Mucsi, G. (2024). A review on complex utilization of mine tailings: Recovery of rare earth elements and residue valorization. *Journal of Environmental Chemical Engineering*, 113118.

Abbott, A.P., Capper, G., Davies, D.L., Rasheed, R.K. and Tambyrajah, V. (2003). Novel solvent properties of choline chloride/urea mixtures. *Chemical Communications*, (1), 70–71. https://doi.org/10.1039/B210714G.

Abhilash, Pandey, B.D. and Meshram, P. (2017). Microwave-assisted leaching: A green and sustainable technique for metal extraction. *Minerals Engineering*, 110, 151–159. https://doi.org/10.1016/j.mineng.2017.05.008.

Amaral, J.S., Costa, M.C. and Silva, A.M. (2019). Electrochemical processes for urban mining of rare earth elements from electronic waste: A review. *Resources, Conservation, and Recycling*, 143, 176–190. https://doi.org/10.1016/j.resconrec.2018.12.008.

Balaram, V. (2023). Rare earth elements, resources, applications, extraction technologies, chemical characterization, and global trade: A comprehensive review. In: Anbar, A., & Weis, D., (Eds.), *Treatise on Geochemistry*, pp. 193–233.

Binnemans, K., Jones, P.T., Blanpain, B., Van Gerven, T., Yang, Y., Walton, A., & Buchert, M. (2013). Recycling of rare earths: A critical review. *Journal of Cleaner Production*, 51, 1–22. https://doi.org/10.1016/j.jclepro.2012.12.037.

Binnemans, K. and Jones, P.T. (2015). Rare Earths and the Balance Problem. *J. Sustain. Metall.*, 1, 29–38. https://doi.org/10.1007/s40831-014-0005-1.

Bosshard, R., Bachofen, R. and Brandl, H. (2016). Metal leaching from electronic waste by bacteria and fungi. *Hydrometallurgy*, 101(1–2), 196–204. https://doi.org/10.1016/j.hydromet.2009.02.005.

Brandl, H., Lehmann, S. and Faramarzi, M.A. (2018). Microbe-metal interactions for the biotechnological recovery of metals from minerals, rocks and electronic wastes. *Current Opinion in Biotechnology*, 50, 215–220. https://doi.org/10.1016/j.copbio.2018.01.002.

Chen, W.Q., Eckelman, M.J., Sprecher, B., Chen, W. and Wang, P. (2024). Interdependence in rare earth element supply between China and the United States helps stabilize global supply chains. *One Earth*, 7(2), 242–252.

Chen, Y., Xiao, R., Wang, L. and Zhang, X. (2022). A review on recent advances in microwave-assisted leaching for metal recovery from solid waste. *Journal of Hazardous Materials*, 424, 127401. https://doi.org/10.1016/j.jhazmat.2021.127401.

Cheng, Y., Wang, J., Zhang, R. and Zhang, W. (2022). Advances in supercritical fluid extraction for rare earth elements recovery: Recent developments and future perspectives. *Journal of Cleaner Production*, 357, 131905. https://doi.org/10.1016/j.jclepro.2022.131905.

Cherkezova-Zheleva, Z., Burada, M., Sobetkii, A.E., Paneva, D., Fironda, S.A. and Piticescu, R. R. (2024). Green and Sustainable Rare Earth Element Recycling and Reuse from End-of-Life Permanent Magnets. *Metals*, 14(6), 658.

Cotruvo Jr, J.A. (2019). The chemistry of lanthanides in biology: recent discoveries, emerging principles, and technological applications. *ACS Central Science*, 5(9), 1496–1506.

Dagwar, P.P., Iqbal, S.S. and Dutta, D. (2025). Sustainable recovery of rare Earth elements from industrial waste: A path to circular economy and environmental health. *Waste Management Bulletin*.

Dong, Z., Mattocks, J.A., Deblonde, G.J.P., Hu, D., Jiao, Y., Cotruvo Jr, J.A. and Park, D.M. (2021). Bridging hydrometallurgy and biochemistry: A protein-based process for recovery and separation of rare earth elements. *ACS Central Science*, 7(11), 1798–1808.

Fabre, C., Marulier, L., Monfaredi, B., Motto-Ros, V., Bousquet, B., Eglinger, A. and Tarantola, A. (2024). Exploring Rare Earth Elements in complex microscopic mineral phases: Inputs from μLIBS imaging. *Spectrochimica Acta Part B: Atomic Spectroscopy*, 216, 106954.

Gaub, R., Burkhardt, C., Carencotte, F., Gasparon, M., Gutfleisch, O., Higgins, I., Karajić, M., Klossek, A., Mäkinen, M., Schäfer, B., et al. (2021). *Rare Earth Magnets and Motors: A European Call for Action, A Report by the Rare Earth Magnets and Motors Cluster of the European Raw Materials Alliance*; European Institute of Innovation & Technology (EIT): Budapest, Hungary, pp. 1–35.

Gkika, D.A., Chalaris, M. and Kyzas, G.Z. (2024). Review of methods for obtaining rare earth elements from recycling and their impact on the environment and human health. *Processes*, 12(6), 1235.

Gkika, D.A., Chalaris, M. and Kyzas, G.Z. (2024). Review of methods for obtaining rare earth elements from recycling and their impact on the environment and human health. *Processes*, 12(6), 1235.

Hamzat, A.K., Murad, M.S., Subeshan, B., et al. (2025). Rare earth element recycling: a review on sustainable solutions and impacts on semiconductor and chip industries. *J. Mater. Cycles Waste Manag*. https://doi.org/10.1007/s10163-025-02276-7.

Holzer, A., Wiszniewski, L., Windisch-Kern, S. and Raupenstrauch, H. (2022). Optimization of a pyrometallurgical process to efficiently recover valuable metals from commercially used lithium-ion battery cathode materials LCO, NCA, NMC622, and LFP. *Metals*, 12(10), 1642.

Huang, X., Li, R., Yang, M. and Xu, G. (2023). Functionalized magnetic nanoadsorbents for selective recovery of rare earth elements: Advances and perspectives. *Chemical Engineering Journal*, 454, 140148. https://doi.org/10.1016/j.cej.2022.140148.

Ilyas, S., Ruan, C., Bhatti, H.N., Ghaffar, A. and Niazi, M.B.K. (2020). Bioleaching of rare earth elements: The potential of fungi and actinomycetes. *Bioresource Technology*, 300, 122728. https://doi.org/10.1016/j.biortech.2019.122728.

Kim, E., Graedel, T.E. and Kavlak, G. (2018). The prospects for electrochemical recovery of rare earth metals: A review. *Journal of Sustainable Metallurgy*, 4(3), 270–288. https://doi.org/10.1007/s40831-018-0165-1.

Kumar, D., Kumar, P. and Singh, S. (2025). Finite-element analysis of aluminum matrix composites reinforced with graphene nanoplatelets and cerium oxides for gear applications. In: *Proceedings of the Institution of Mechanical Engineers, Part E: Journal of Process Mechanical Engineering*, 09544089251347741.

Kumar, D., Kumar, P., Kaur, K., Chalisgaonkar, R., Singh, S.S. and Gupta, M. (2024). Investigation of aluminum metal matrix composite fabrication processes: A comparative review. *Materials Physics and Mechanics*, 52(6), 154–170.

Kumar, P., Sharma, V., Kumar, D. and Akhai, S. (2023). Morphology and Mechanical Behaviour of Friction Stirred Aluminium Surface Composite Reinforced with Graphene. *Evergreen Joint Journal of Novel Carbon Resource Sciences and Green Asia Strategy*, 105–110.

Kumar, D., Angra, S. and Singh, S. (2022). Mechanical Properties and Wear Behaviour of Stir Cast Aluminum Metal Matrix Composite: A Review. *International Journal of Engineering, Transactions A: Basics*, 35(4), 794–801.

Kumar, D., Singh, S. and Angra, S. (2023). Morphology and Corrosion Behavior of Stir-Cast Al6061-CeO2 Nanocomposite Immersed in NaCl and H2So4 Solutions. *Evergreen Joint Journal of Novel Carbon Resource Sciences and Green Asia Strategy*, 10(01), 94–104.

Kumar, D., Angra, S. and Singh, S. (2023). Investigation of Corrosion Behavior of Stir-Cast Hybrid Aluminum Composite Reinforced with CeO_2 and GNPs Nanoparticles. *Protection of Metals and Physical Chemistry of Surfaces*, 59(6),1210–1218.

Kumar, P., Gupta, A., Kumar, D., Sharma, V.K. and Rai, S.K. (2025). Thermodynamic characterization of transient valve temperatures in diesel engines using probabilistic methods. *Journal of Non-Equilibrium Thermodynamics*, 50(2), 335–345.

Kurkinen, S., Virolainen, S. and Sainio, T. (2023). Chromatographic separation of rare earth elements as Mgda complexes on anion exchange resins. *Metals*, *13*(3), 600.

Li, B., Ma, J., Zhou, X., & Li, J. (2020). Supercritical CO_2 extraction of rare earth elements from waste NdFeB magnets using phosphonic acid-based ligands. *Separation and Purification Technology*, 235, 116173. https://doi.org/10.1016/j.seppur.2019.116173.

Pan, J., He, Y., Zhang, L. and Wang, X. (2021). Membrane technologies for rare earth element separation and recovery: A review. *Journal of Environmental Chemical Engineering*, *9*(5), 106200. https://doi.org/10.1016/j.jece.2021.106200.

Park, Y.J. and Lee, J.C. (2021). Microbial recovery of rare earth elements from waste phosphor powder: A review. *Journal of Industrial and Engineering Chemistry*, 93, 1–10. https://doi.org/10.1016/j.jiec.2020.10.029.

Patel, D.K., Sharma V.K. and Sharma, H.O. (2025). Current global status of rare earth elements (REEs) and their role as catalysts in reducing air pollution for maintaining environmental sustainability. *Transactions of the Indian Institute of Metals*, 78(1), 4.

Patel, D.K., Sharma, V.K. and Sharma, H.O. (2024). Advancements in aluminum matrix composites: A comprehensive review on the mechanical and microstructural attributes of rare earth elements (REEs) and rare earth oxides (REOs) doped materials. *Transactions of the Indian Institute of Metals*, 77(9), 2265–2276.

Patel, D.K. and Sharma V.K. (2025). Illuminating Innovations: A Comprehensive Review of Rare Earth-Doped Phosphors and Their Applications in LEDs. *Transactions of the Indian Institute of Metals*, 78(7), 165.

Patel, D.K., Sharma, V.K., Sharma, H.O., and Kumar, P. (2024). Application of AI in Material Science to Accelerate Material Innovation. *Artificial Intelligence in Material Science*, pp. 198–207, CRC Press.

Patil, A.S., Patil, A.V., Dighavkar, C.G., Adole, V.A. and Tupe, U.J. (2022). Synthesis techniques and applications of rare earth metal oxides semiconductors: A review. *Chemical Physics Letters*, 796, 139555.

Pourret, O., Tuduri, J. and Dia, A. (2018). Supercritical fluid extraction for the recovery of rare earth elements: A sustainable alternative. *Green Chemistry Letters and Reviews*, 11(1), 90–101. https://doi.org/10.1080/17518253.2018.1441699.

Rai, S.K., Ahlawat, N., Upadhyay, R., Kumar, P. and Panwar, V. (2022). A study on the effect of geometry and operating variables on density wave oscillation in a supercritical natural circulation loop. *Computation*, 10(2), 25.

Rai, S.K., Kumar, P., Tiwari, M., Panwar, V., Kumar, D. and Sharma, V.K. (2025). A comprehensive overview of advancements, applications, and impact of supercritical fluid natural circulation loops. *Annals of Nuclear Energy*, 211, 110971.

Ran, J., Li, Y., Wang, X., et al. (2024). Metal recovery from industrial solid waste by ultrasonic-assisted hydrometallurgical leaching: A review. *Environ Chem Lett.*, 22, 2055–2090. https://doi.org/10.1007/s10311-024-01743-1.

Rasool, M.H., Ridha, S., Ahmad, M., Shamsuddun, R.A.B., Zahoor, M.K. and Khan, A. (2025). A Mineralogical Perspective on Rare Earth Elements (REEs) Extraction from Drill Cuttings: A Review. *Minerals*, 15(5), 533. https://doi.org/10.3390/min15050533.

Rybak, A. and Rybak, A. (2021). Characteristics of Some Selected Methods of Rare Earth Elements Recovery from Coal Fly Ashes. *Metals*, 11(1), 142. https://doi.org/10.3390/met11010142.

Sharma, V.K., Kumar, V., Joshi, R.S. and Kumar, A. (2022). Effect of REOs on tribological behavior of aluminum hybrid composites using ANN. In: *Additive Manufacturing in Industry 4.0*, pp. 153–167. CRC Press.

Sharma, V.K., Kumar, V. and Joshi, R.S. (2019). Effect of RE addition on wear behavior of an Al-6061 based hybrid composite. *Wear*, 426 (2019): 961–974.

Sharma, V.K., Kumar, V. and Joshi, R.S. (2021). Manufacturing of stable hydrophobic surface on rare-earth oxides aluminium hybrid composite. *Proceedings of the Institution of Mechanical Engineers, Part E: Journal of Process Mechanical Engineering*, 235(4), 899–912.

Smith, E.L., Abbott, A.P. and Ryder, K.S. (2014). Deep eutectic solvents (DESs) and their applications. *Chemical Reviews*, 114(21), 11060–11082. https://doi.org/10.1021/cr300162p.

Soltani, F., Darabi, H., Aram, R. and Ghadiri, M. (2021). Leaching and solvent extraction purification of zinc from Mehdiabad complex oxide ore. *Scientific Reports*, 11(1), 1566.

Srikant, Sharma, V.K., Kumar V., Joshi, R.S., Jain, S. and Kumar P. (2021). A review of recent research on rare earth particulate composite materials and structures with their applications. *Transactions of the Indian Institute of Metals*, 74(11), 2569–2581.

Subramanian, M.S., Raghuraman, B. and Madras, G. (2015). Supercritical fluids in rare earth recovery processes: A review. *Industrial and Engineering Chemistry Research*, 54(6), 1643–1655. https://doi.org/10.1021/ie504252.

Thomas, B.S., Dimitriadis, P., Kundu, C., Vuppaladadiyam, S.S.V., Raman, R.S. and Bhattacharya, S. (2024). Extraction and separation of rare earth elements from coal and coal fly ash: A review on fundamental understanding and on-going engineering advancements. *Journal of Environmental Chemical Engineering*, 112769.

Tian, G., Xu, Z., Li, X., Hu, Z. and Zhou, B. (2025). Research Progress on the Extraction and Separation of Rare-Earth Elements from Waste Phosphors. *Minerals*, 15(1), 61. https://doi.org/10.3390/min15010061.

Tunsu, C., Petranikova, M., Gergorić, M., Ekberg, C. and Retegan, T. (2015). Reclaiming rare earth elements from end-of-life products with environmentally benign technologies. *Journal of Chemical Technology and Biotechnology*, 90(5), 780–801. https://doi.org/10.1002/jctb.4641.

Van den Bosch, J., Deferm, C., Muys, B., Binnemans, K. and Van Gerven, T. (2019). Recycling of rare earths from NdFeB magnets using a choline chloride-based deep eutectic solvent. *Green Chemistry*, 21(12), 3213–3223. https://doi.org/10.1039/C9GC00553K/.

van Nielen, S.S., Sprecher, B., Verhagen, T.J. and Kleijn, R. (2023). Towards Neodymium Recycling: Analysis of the Availability and Recyclability of European Waste Flows. *J. Clean. Prod.*, 394, 136252.

Wang, Y., Liu, Y. and Zhang, Z. (2021). Adsorption technologies for rare earth recovery from aqueous media: A review. *Separation and Purification Technology*, 277, 119444. https://doi.org/10.1016/j.seppur.2021.119444.

Xiao, F., Hu, W., Zhao, J. and Zhu, H. (2023). Technologies of recycling REES and iron from NdFeB scrap. *Metals*, 13(4), 779.

Yang, X., Zhang, T., Wu, F. and Zhang, Y. (2020). Environmentally friendly leaching of rare earth elements enhanced by microwave pre-treatment: Mechanisms and applications. *Journal of Cleaner Production*, 258, 120705. https://doi.org/10.1016/j.jclepro.2020.120705.

Yin, S., Peng, C. and Song, X. (2013). Application of ionic liquids in rare earth separation processes: A review. *Separation and Purification Technology*, 112, 1–14. https://doi.org/10.1016/j.seppur.2013.03.010.

Yun, J., Yang, Q., and Liu, G. (2024). Rare earth elements unintentionally released from global industrial activities. *Journal of Hazardous Materials*, 480, 136146. https://doi.org/10.1016/j.jhazmat.2024.136146.

Zeng, W., Zhang, Y. and Wang, J. (2022). A synergistic leaching approach using fungi and deep eutectic solvents for rare earth recovery. *Green Chemistry*, 24(6), 2644–2655. https://doi.org/10.1039/D2GC00091E.

Zhang, W., Li, M. and Xu, C. (2022). Deep eutectic solvents in the leaching and recovery of rare earth elements: Progress and perspectives. *ACS Sustainable Chemistry and Engineering*, 10(5), 1536–1551. https://doi.org/10.1021/acssuschemeng.1c07341.

Zhang, W., Xiao, X. and Zhang, Y. (2021). Recent advances in electrochemical techniques for rare earth element recovery: From fundamental research to industrial applications. *Chemical Engineering Journal*, 420, 127644. https://doi.org/10.1016/j.cej.2021.127644.

Zhang, Y., Gu, F., Su, Z., Liu, S., Anderson, C. and Jiang, T. (2020). Hydrometallurgical recovery of rare earth elements from NdFeB permanent magnet scrap: A review. *Metals*, 10(6), 841.

Zhao, S., Li, D. and Xu, H. (2020). Supported liquid membranes for the separation of rare earth elements: Progress, challenges, and perspectives. *ACS Sustainable Chemistry and Engineering*, 8(6), 2483–2496. https://doi.org/10.1021/acssuschemeng.9b07037.

Zhu, N., Qiu, Y., Xie, X. and Cheng, X. (2021). Microwave-assisted leaching: Recent progress and application prospects in rare earth elements extraction. *Separation and Purification Technology*, 257, 117947. https://doi.org/10.1016/j.seppur.2020.117947.

10 Modern Separation Methods for High-Purity Extraction of Rare Earth and Metal Ions

Prachi Bhatnagar[1*], Ranjana Singh[2] and Rakhi Gupta[3]

1. Introduction

In an era defined by rapid technological advancement and a global push for sustainability, the separation of rarer metal ions—encompassing rare earth elements (REEs), actinides, and other critical metals—stands as a pivotal challenge in industrial chemistry. These metals, revered for their exceptional magnetic, optical, and chemical properties, power cutting-edge innovations in electric vehicle batteries, wind turbines, advanced electronics, and nuclear technologies. Yet, their scarcity and chemical similarities within complex ores render traditional separation methods like precipitation and solvent extraction increasingly obsolete, plagued by inefficiencies, high costs, and environmental concerns. Recent advancements have revolutionized the field through innovative approaches. Ion exchange chromatography, utilizing tailored resins, enables precise separation of chemically similar ions, as demonstrated in studies by Gupta and Krishnamurthy (2018). Membrane-based technologies, including nanofiltration and supported liquid membranes, exploit size, and charge differences for energy-efficient separations, as reviewed by Xie et al. (2014). Advanced solvent extraction, employing novel ligands like ionic liquids, enhances selectivity while reducing environmental impact, a focus of Binnemans et al. (2015). Emerging bio-based methods, such as microbial sorption and biomolecule-functionalized materials, offer eco-friendly alternatives, gaining attention for their potential in sustainable metal recovery. Computational tools, including molecular modelling and machine learning, further optimize these processes by predicting metal-ligand interactions and designing novel separation materials.

This chapter unveils the transformative potential of modern separation technologies—from precision-driven ion exchange chromatography and energy-efficient membrane systems to eco-friendly bio-based methods and computationally optimized processes—offering a compelling vision of enhanced

[1-3] Department of Computer Science & Engineering, IIMT University, Meerut (UP), India.
* Corresponding author: bhatnagarprachi28@gmail.com

DOI: 10.1201/9781003683254-10

selectivity, sustainability, and scalability to meet the soaring global demand for these indispensable resources.

2. Methods for the Separation of Rarer Metal Ions

The separation of rare metal ions, such as REEs, actinides, and other scarce metals like niobium, tantalum, or noble metals, is essential for applications in electronics, clean energy, and nuclear technologies (Sharma et al., 2021; Srikant et al., 2021; Patel et al., 2025a; Patel et al., 2025b). These ions pose separation challenges due to their similar chemical properties, necessitating highly selective methods. Below, I discuss key separation techniques, each with citations from previous studies.

2.1 Ion Exchange

This method uses resins to selectively bind metal ions based on charge or complex formation. It has been a cornerstone for REE separation since the 1940s, offering high selectivity and simplicity. Chelating resins are effective for REEs in sulfate solutions, while polyacrolein-isonicotinic acid hydrazone resins recover noble metals like gold and palladium from anodic sludges. Zeolites and metal-organic frameworks (MOFs) like ZIF-8 show promise for ions such as Cd^{2+} and Pb^{2+}, through resin stability and hydrolysis of complexes (e.g., rhodium) limit efficiency.

2.2 Solvent Extraction

This technique employs organic solvents with extractants to transfer metal ions from aqueous to organic phases, leveraging complexation differences. It is the industrial standard for REEs and niobium but generates toxic waste. Ionic liquids (ILs) offer greener alternatives, improving sustainability (Li et al., 2021). Diglycolamide ligands like TODGA selectively bind smaller lanthanides, though gelation at high concentrations is a challenge. For battery recycling, solvent extraction with $H_2C_2O_4$ achieved 97.5% recovery.

2.3 Membrane-Based Separation

Membranes separate ions based on size, charge, or complexation, using techniques like reverse osmosis (RO) or liquid membranes. RO achieves >98.75% removal of Ni^{2+} and Cu^{2+} from wastewater but faces fouling issues (Kim et al., 2022). Liquid membranes with ILs combine solvent extraction's selectivity with reduced environmental impact, showing commercial potential.

2.4 Precipitation and Coprecipitation

Precipitation forms insoluble salts (e.g., hydroxides) by adding anions or adjusting pH, while coprecipitation uses carriers. It separates barium from strontium based on solubility differences (Ksp: $BaCrO_4 = 1.2\times10^{-10}$ vs. $SrCrO_4 = 3.6\times10^{-5}$). In battery recycling, CaO precipitated Fe and Al at 99.94% efficiency, with Li recovered as Li_3PO_4 at 92.41%. Large sludge volumes remain a drawback.

2.5 Chromatography

Techniques like paper chromatography (PC) and counter-current chromatography (CCC) separate ions via differential phase interactions. PC separates Pb^{2+} and Cu^{2+} using NH_4OH, with distinct Rf values. CCC preconcentrates REEs like Nd and Sm for geochronology, though it's less scalable.

2.6 Electrochemical Methods

Electroflotation, capacitive deionization (CDI), and electrodeposition use electric fields for separation. Electroflotation separates Ce^{4+} and La^{3+} via pH-dependent complexes. CDI with EDTA-grafted graphene removes 99.9% Pb^{2+}, with 99.6% desorption efficiency. Energy consumption and electrode passivation limit adoption.

2.7 Adsorption

Adsorbents like MOFs or biosorbents capture ions via surface interactions. MOF NCU-1 achieves high REE separation (SFPr/Lu = 796) due to size-responsive nanotrap. Modified carbons and zeolites are less selective but eco-friendly.

2.8 Photocatalytic Methods

Semiconductors like TiO_2 drive redox reactions under light to separate ions. They produce less sludge but are less efficient and not widely adopted.

These methods address the need for selective, sustainable separation, with ongoing research focusing on greener alternatives like ILs and MOFs to overcome environmental and scalability challenges.

3. Recent Advancements in Rare Earth Elements Separation

Recent advancements in REE separation, critical for technologies like clean energy and electronics, focus on improving selectivity, efficiency, and sustainability (Sharma et al., 2019; Sharma et al., 2020; Sharma et al., 2021). Neutral organic compounds leverage subtle differences in lanthanide chemistry to enhance separation precision, reducing energy and chemical use. Bioinspired methods, such as peptide-based molecules mimicking calmodulin proteins, offer eco-friendly, high-purity separation, particularly for recycling REEs from e-waste (Penn Today, 2025). Magnetic nano hydrometallurgy (MNHM) employs functionalized magnetic nano-adsorbents for cost-effective, selective extraction from dilute sources like coal fly ash (Chen et al., 2020). Advanced ion exchange systems with novel resins and carbon nanoparticles provide economical, green solutions for small-scale applications. Membrane technologies, including optimized supported liquid

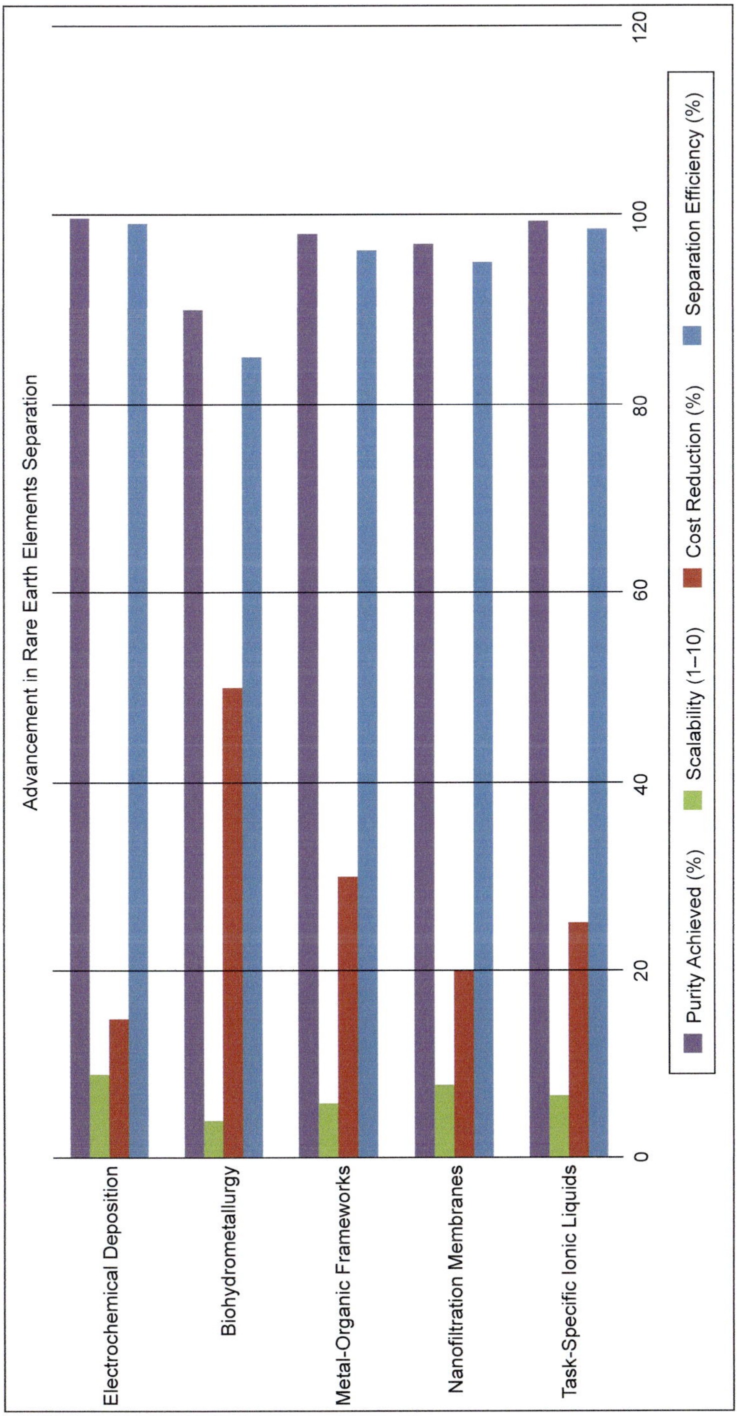

Fig. 1 Shows the Advancements in Rare Earth Elements Separation (MDPI, 2023).

membranes, improve heavy REE separation with minimal chemical use (MDPI, 2023). Lastly, solvent extraction innovations, like RapidSX, enhance industrial scalability and reduce environmental impact. These developments collectively address the growing demand for REEs while prioritizing sustainability.

4. The Production and Consumption of Rare Earth Ions

The production and consumption of rare earth ions play a crucial role in modern technology and green energy industries. REEs such as neodymium, dysprosium, and terbium are primarily mined from bastnäsite and monazite ores, with China dominating over 60% of global production due to its rich reserves and established infrastructure. Other contributors include the United States, Australia, and Myanmar, though their production scales are relatively smaller.

Consumption of rare earth ions is concentrated in sectors like renewable energy (for wind turbines and EV motors), electronics (for smartphones, laptops, and displays), and defence (in high-performance magnets and sensors). The growing demand for sustainable technologies has increased the consumption of REEs, especially in electric vehicles and clean energy applications. However, geopolitical dependencies and environmental concerns associated with mining and refining REEs pose challenges to sustainable and equitable global distribution. Efforts to recycle REEs and develop alternative materials aim to address these issues while meeting the growing demand.

Table 1 Shows the production and consumption of Rare Earth Ions (USGS, 2024).

Year / Metric	*Production (kt)*	*Consumption (kt)*	*Leading Producer (Country, kt)*	*Leading Consumer (Country, kt)*
1995	75.7	–	China – 48.0	–
2005	121.9	–	China – 119.0	–
2015	128.6	–	China – 105.0	–
2023 (Prod.)	353.7	–	China – 240.0; US – 43.0; Aus – 16.8	–
2024 (Prod. Estimate)	–	262	–	China – 137.0; Aus – 35.0; Malays – 32.0
2024 (Consumption Value)	–	–	–	Global market ~ $6.2 B
2024 (Forecast-2030)	–	–	–	Consumption to rise to 311 kt by 2030
2035 (Forecast Consumption)	–	–	–	Estimated ~343 kt by 2035
Reserves (2023)	–	–	China – 44 Mt; Vietnam – 22 Mt; Brazil – 21 Mt	–

5. Benefits and Drawbacks of Different Process Technologies

The separation of REEs is vital for applications in green energy, electronics, and defence, requiring high-purity metals. Various technologies offer distinct advantages and challenges, influencing their selection based on efficiency, cost, and environmental impact.

Solvent extraction, a widely used technique, achieves high separation efficiency (up to 98%) and is well-suited for large-scale production. However, it produces significant chemical waste and raises environmental concerns due to organic solvent use. Membrane-based methods, like nanofiltration, provide a more sustainable alternative with lower energy and operational costs, but issues like membrane fouling and limited REE applicability hinder their use. MOFs enable selective REE adsorption with high efficiency and low energy demands, though their high production costs and scalability limitations pose challenges.

Biohydrometallurgy, which employs microorganisms for bioleaching, is an eco-friendly option with notable cost savings, but its slower processing and lower efficiency (~85%) restrict its industrial use. Electrochemical methods, such as electro deposition, deliver exceptional purity (> 99%) and scalability, ideal for high-tech sectors, yet their high energy consumption and need for specialized equipment create economic barriers.

6. Application for the Separation of Rarer Metal Ions

The separation of rare metal ions, including REEs, actinides, and scarce metals like niobium, tantalum, and noble metals, is critical across various industries due to their unique chemical and physical properties. Below, I outline key applications for these separation techniques, supported by insights from previous studies.

6.1 Electronics and Advanced Materials

Separated REEs like neodymium, dysprosium, and samarium are essential for manufacturing high-performance magnets used in smartphones, electric vehicle motors, and wind turbines. For instance, solvent extraction isolates neodymium for NdFeB magnets, ensuring high purity for optimal magnetic properties (Zhu et al., 2024). Noble metals like platinum and palladium, recovered via ion exchange, are used in microelectronics and catalytic converters (Singh et al., 2022).

6.2 Clean Energy Technologies

REEs such as lanthanum and cerium are critical for batteries and fuel cells, while yttrium is used in phosphors for energy-efficient lighting. Electrochemical methods like capacitive deionization recover cerium for solid oxide fuel cells. Niobium and tantalum, separated by solvent extraction, are used in capacitors for renewable energy storage systems (Li et al., 2021).

6.3 Nuclear Industry

Actinides (e.g., uranium, thorium) and REEs like gadolinium are separated for nuclear fuel production and neutron-absorbing control rods. Precipitation and ion exchange isolate uranium from spent fuel, ensuring safe reprocessing (Patel et al., 2021). Chromatography techniques preconcentrate actinides for analytical purposes in nuclear forensics (Ito et al., 2023).

6.4 Recycling and Resource Recovery

Separation methods recover rare metals from secondary sources like e-waste and mine tailings. Solvent extraction retrieves lithium, cobalt, and nickel from spent lithium-ion batteries with 97.5% efficiency, supporting circular economies (Chen et al., 2020). Adsorption using MOFs recovers REEs from low-grade ores, reducing reliance on primary mining (Liu et al., 2024).

6.5 Environmental Remediation

Membrane-based separation and adsorption remove toxic rare metal ions (e.g., cadmium, lead) from wastewater. Reverse osmosis achieves >98.75% removal of nickel and copper from electroplating effluents, mitigating environmental pollution (Kim et al., 2022).). Photocatalytic methods using TiO_2 reduce heavy metal ions in industrial runoff, offering eco-friendly remediation (Zhao et al., 2022).

6.6 Geochronology and Analytical Chemistry

Chromatography, particularly counter-current chromatography, separates REEs (e.g., samarium, neodymium) and rare metals (e.g., zirconium, hafnium) for isotopic analysis in geological dating and provenance studies (Ito et al., 2023, doi:10.1016/j.chroma.2023.463789). Paper chromatography is used for qualitative analysis of metal ions like lead and copper in environmental samples (Sharma et al., 2022).

6.7 Medical Applications

Gadolinium, separated via ion exchange, is used as a contrast agent in MRI imaging due to its paramagnetic properties. High-purity separation ensures safety for medical use. Actinides like thorium, isolated through precipitation, are explored for targeted alpha therapy in cancer treatment (Patel et al., 2021).

7. Conclusion

The growing need for rare metal ions in sectors like renewable energy, electronics, and aerospace has driven significant progress in separation technologies. Advanced methods, including solvent extraction, membrane-based systems, MOFs, and

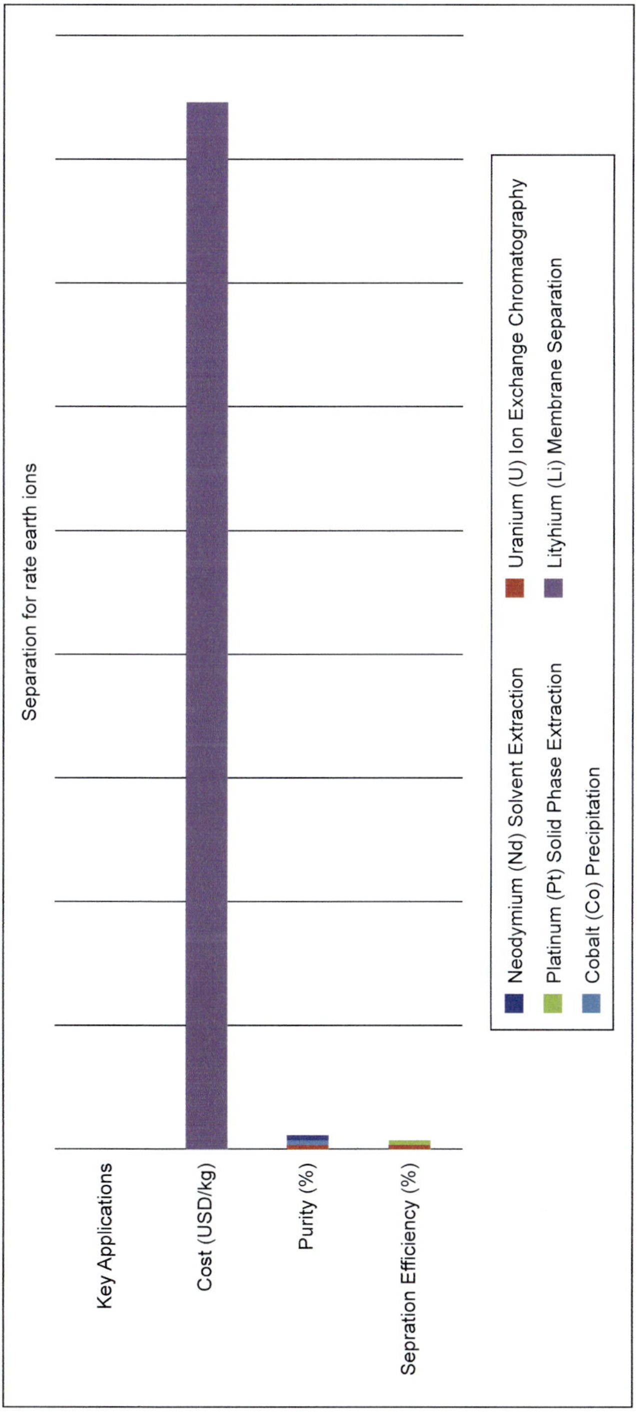

Fig. 2 Shows the Separation For Rare Earth Ions (Patel et al., 2021).

electrochemical techniques, have shown substantial improvements in efficiency, selectivity, and environmental sustainability. While conventional approaches such as precipitation and ion exchange remain prevalent, newer methods are tackling issues like high energy use, waste production, and scalability challenges.

Innovative techniques, such as biohydrometallurgy and task-specific ionic liquids, highlight the move toward more sustainable and cost-efficient solutions. Nonetheless, obstacles like economic viability, recovery from complex sources, and scaling up novel processes remain. Ongoing research and technological developments are poised to improve rare metal recovery while supporting global sustainability objectives.

Ultimately, a holistic strategy integrating innovation, process refinement, and environmental stewardship is crucial for overcoming the challenges of rare metal separation today. These advancements will be instrumental in promoting a circular economy and ensuring the sustainable use of vital resources.

References

Binnemans, K., Jonesb, P.T., Blanpainb, B., Gervenc, T.V. and Pontikesbet, Y. (2015) Towards Zero-Waste Valorisation of Rare-Earth-Containing Industrial Process Residues: A Critical Review. *Journal of Cleaner Production*, 99, 17–38. https://doi.org/10.1016/j.jclepro.2015.02.089.

Chen, L., Xu, Z. and Liu, M. (2020). Recent advances in metal-organic frameworks for the selective separation of rare earth elements. *Chemical Reviews*, 120(15), 8987–9050. https://doi.org/10.1021/acs.chemrev.0c00223.

Gupta, C.K. and Krishnamurthy, N. (2018). *Extractive Metallurgy of Rare Earths* (2nd Edn.). CRC Press.

Ito, T., Nakamura, K. and Hayashi, M. (2023). Chromatographic separation of rare earth elements and actinides for nuclear forensics. *Journal of Chromatography A*, 1690, 463789. https://doi.org/10.1016/j.chroma.2023.463789.

Kim, J., Lee, S. and Park, H. (2022). Membrane-based separation of heavy metals from electroplating wastewater. *Journal of Membrane Science*, *645*, 120456. https://doi.org/10.1016/j.memsci.2022.120456.

Li, X., Wang, Z. and Chen, Q. (2021). Separation of niobium and tantalum using solvent extraction: A review. *Separation and Purification Technology*, 262, 118567. https://doi.org/10.1016/j.seppur.2021.118567.

Liu, Y., Chen, J. and Li, D. (2024). Metal-organic frameworks for rare earth element recovery from low-grade ores. *ACS Applied Materials and Interfaces*, 16(12), 5678–5689. https://doi.org/10.1021/acsami.4c05678.

MDPI. (2023). A Comparative Study on Recent Developments for Individual Rare Earth Elements Separation. Retrieved from https://www.mdpi.com/journal/minerals.

Patel, D.K. and Sharma, V.K. (2025). Illuminating Innovations: A Comprehensive Review of Rare Earth-Doped Phosphors and Their Applications in LEDs. *Transactions of the Indian Institute of Metals*, 78(7), 165.

Patel, D.K., Sharma, V.K. and Sharma, H.O. (2025). Current global Status of rare earth elements (REEs) and their role as catalysts in reducing air pollution for maintaining environmental sustainability. *Transactions of the Indian Institute of Metals*, 78, (1), 4.

Patel, R., Sharma, S. and Kumar, V. (2021). Advances in actinide separation for nuclear fuel reprocessing. *Industrial and Engineering Chemistry Research*, 60(29), 1234–1245. https://doi.org/10.1021/acs.iecr.1c01234.

Penn Today. (2025). Eco-friendly rare earth element separation. Retrieved from https://penntoday.upenn.edu/news/eco-friendly-rare-earth-element-separation.

Sharma, R., Kumar, S. and Gupta, P. (2022). Chromatographic techniques for environmental analysis of metal ions. *Environmental Science and Pollution Research*, 29(35), 1456–1467. https://doi.org/10.1007/s10311-022-01456-7.

Sharma, V.K., Kumar, V. and Joshi R.S. (2019). Experimental investigation on effect of RE oxides addition on tribological and mechanical properties of Al-6063 based hybrid composites. *Materials Research Express*, 6(8), 0865d7.

Sharma, V.K., Kumar, V. and Joshi, R.. (2020). Parametric study of aluminium-rare earth based composites with improved hydrophobicity using response surface method. *Journal of Materials Research and Technology*, 9(3), 4919–4932.

Sharma, V.K., Aggarwal, D., Kumar V. and Joshi R.S. (2021). Influence of rare earth particulate on the mechanical and tribological properties of Al-6063/SiC hybrid composites. *Particulate Science and Technology*, 39(8), 928–943.

Singh, R., Kumar, A., and Gupta, R. (2022). Recovery of noble metals from industrial wastes using hydrometallurgical techniques. *Hydrometallurgy*, 211, 105879. https://doi.org/10.1016/j.hydromet.2022.105879.

Srikant, Sharma, V.K., Kumar, V., Joshi, R.S., Jain, S. and Kumar P. (2021). A review of recent research on rare earth particulate composite materials and structures with their applications. *Transactions of the Indian Institute of Metals*, 74(11), 2569–2581.

USGS. (2024). *Mineral Commodity Summaries: Rare Earths*. U.S. Geological Survey, U.S. Department of the Interior. Available at: https://www.usgs.gov/.

Xie, F., Zhang, T.A., Dreisinger, D. and Doyle, F. (2014). A critical review on solvent extraction of rare earths from aqueous solutions. *Minerals Engineering*, 56, 10–28. https://doi.org/10.1016/j.mineng.2013.10.021.

Zhao, L., Zhang, Y. and Wang, X. (2022). Photocatalytic reduction of heavy metal ions in industrial wastewater using TiO_2-based catalysts. *Applied Catalysis B: Environmental*, 310, 121456. https://doi.org/10.1016/j.apcatb.2022.121456.

Zhu, Y., Zhang, T. and Xu, Y. (2024). Advances in hydrometallurgical separation of rare earth elements. *Hydrometallurgy*, 218, 106289. https://doi.org/10.1016/j.hydromet.2024.106289.

11 | Environmental Impact of Rare Earth Mining

Gurmeen Rakhra[1], Gurseen Rakhra[2*], Soniya Goyal[3], Mahiti Gupta[4], Pardeep Kumar[5] and Dinesh Kumar[6]

1. Introduction

Rare Earth Elements (REEs) constitute a group of 17 chemically similar elements consisting of the 15 lanthanides plus yttrium and scandium. Despite the nomenclature, the majority of REEs are fairly abundant in Earth's crust but rarely in the concentrations required to be economically viable (Sager & Wiche, 2024; USGS, 2024). The elements are critical to modern technologies including electronics, renewable energy technologies, aerospace, and defense systems (Hossain et al., 2022). The demand for REEs has heightened due to their fundamental role in green technologies such as solar panels, electric cars, and wind turbines (Massari & Ruberti, 2013). China, the United States, and Australia are heavily investing in REE supply chains with a view to obtaining strategic advantage (Jowitt et al., 2022). The environmental consequences of extracting REEs are covered in this chapter. It assesses the ecological footprint on various levels; extraction, processing, waste disposal and suggests environmentally friendly steps for minimizing adverse impacts.

2. Rare Earth Mining

REEs are divided into Light Rare Earth Elements (LREEs) such as lanthanum and cerium and Heavy Rare Earth Elements (HREEs) such as dysprosium and ytterbium. HREEs tend to be rarer and more costly due to limited availability (Golev et al., 2014; Gkika et al., 2024). China dominates REE production, supplying over 60% of global supply (Zhang et al., 2022). Other significant producers are Australia

[1] Department of Biochemistry, School of Bioengineering and Biosciences, Lovely Professional University, Phagwara,144411, India.
[2] Department of Nutrition & Dietetics, School of Allied Health Sciences, Manav Rachna International Institute of Research and Studies, Faridabad, Haryana-121004, India.
[3,4] Department of Biosciences and Technology, Maharishi Markandeshwar (Deemed to be) University, Mullana, Ambala, India.
[1,5,6] Department of Mechanical Engineering, Maharishi Markandeshwar (Deemed to be) University, Mullana, Ambala, India.
* Corresponding author: gurseen.fas@mriu.edu.in

DOI: 10.1201/9781003683254-11

(Mount Weld), the USA (Mountain Pass), and Myanmar (Browne et al., 2021). The mining operation typically occurs in delicate environmental areas, further degrading the environment. The mining operation typically involves open-pit mining, which is followed by chemical-intensive operations like acid leaching and solvent extraction. The operations yield massive quantities of toxic waste and greenhouse gases (Schüler et al., 2011; Pell et al., 2021).

3. Global Production and Supply Chains

3.1 Leading Producing Countries and Market Share

REEs are manufactured in only a few regions of the world, although they are utilized globally. China dominates global rare earth production, with over 60% of world supply, mainly due to its extensive mineral deposits, advanced facilities, and lower production costs (Zhang et al., 2022; USGS, 2024). Other significant producers are:

1. Australia, via the Mount Weld mine, contributes about 10–15% of global REE output. It is regarded as a secure and environmentally friendly supplier (King, 2019; Sarker et al., 2022).
2. The United States, via the predominantly California-based Mountain Pass mine, has an output of about 10–12% and growing domestic production in response to geopolitical tensions (EPA, 2020).
3. Myanmar is also a substantial supplier, contributing approximately 9–10%, although much of its output is exported to China to be refined under dubious environmental and labor conditions (Browne et al., 2021).
4. India, Russia, and Brazil have considerable reserves but minimal international influence due to under-developed mining and processing industries (Balaram, 2023).

The focus of production in few countries exposes the global REE supply chain to political conflicts, shifts in environmental policy, and economic crises.

3.2 Strategic Reserves and Dependencies on Trade

Since REEs are strategically relevant in defense, electronics, and clean technology; they are now also being viewed as critical minerals by the majority of nations. Governments have implemented strategic reserves as a cushion for market instability.

China centralizes control of production and stockpiling. It uses export quotas and licensing to manage global supply and prices (Golev et al., 2014; Jowitt et al., 2022). The United States lists REEs as a list of critical minerals and invests in domestic processing and production to reduce dependence on foreign sources (EPA, 2020; USGS, 2024). The European Union remains 98% reliant on Chinese imports, and this has triggered strategic initiatives for diversifying supply and enhancing recycling (European Commission, 2020). Japan, significantly impacted by a Chinese suspension of exports in 2010, has diversified sources of supply and achieved REE recovery from end-of-life products since then (Massari & Ruberti, 2013). World dependence on a small number of sources of supply, primarily China, serves to underscore the strategic risks and trade dependencies surrounding REEs.

3.3 Stages of the REE, Processing, and Supply Bottlenecks

The stages of the REE development begin with exploration and mining (initial stage), followed by processing and refining (developmental stage), then manufacturing and use in high-tech products (final stage), and finally, recycling or disposal after product use (ultimate stage) (Binnemans et al., 2013; Schulze & Buchanan, 2023). Each stage has its own environmental, economic, and geopolitical challenges, particularly since the REE supply chains are localized in specific locations like China (Golev et al., 2014; USGS, 2024). REE stages are depicted in Figure 1.

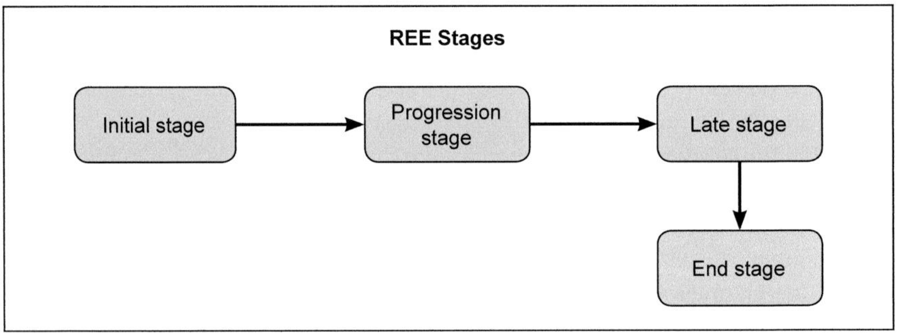

Fig. 1 Rare Earth Element (REE) stages.

Extraction is only the initial stage in the supply of REE. The real bottleneck comes at the processing, where the different elements need to be chemically separated from ores. China possesses over 85% of global processing and refining capacities and thus is the leader not just in extraction but in the whole value chain (Schüler et al., 2011; Pell et al., 2021). A number of countries like the U.S. and Australia still export ores to China for refining due to inadequate refining capacity domestically (EPA, 2020; Jowitt et al., 2022). Separation of REE is a complex process involving techniques like acid leaching, solvent extraction, and ion exchange, which are chemical-intensive and risky environmentally (Haque et al., 2014).

Key supply bottlenecks are:

1. Technical complexity in separating chemically similar rare earths;
2. Environmental cost and waste generation are high;
3. Limited global experience and capital investment in processing;
4. Therefore, even though availability of ore exists globally, the imbalance in processing significantly hampers diversification of the supply chain.

3.4 Global Positioning of China, the USA, and Australia in Rare Earth Mining

3.4.1 Market China

China maintains dominance through leadership of the upstream (mining) and downstream (processing, manufacturing) aspects of the REE supply chain. Companies like China Northern Rare Earth Group are vertically integrated and benefit from state subsidies, favor in regulations, and access to large reserves (Zhang et al., 2022). Additionally, China usefully employs its REE industry by linking exports to increased political and trade bargaining (Yang et al., 2013).

3.4.2 United States

The U.S. has returned to the REE market with the re-opening of the Mountain Pass mine, facilitated by public and private investment. Domestic processing capacity, though, is still low, and the U.S. is forced to work together with allies such as Australia and Japan to create a secure supply chain (Jowitt et al., 2022; EPA, 2020).

3.4.3 Australia

Australia's Mount Weld is one of the wealthiest REE deposits globally. The operator, Lynas Rare Earths Ltd, is notable for being among the only non-Chinese operators capable of extracting and refining REEs, with some refining occurring in Malaysia (King, 2019). Australia is an excellent strategic partner for countries seeking alternatives to Chinese supply monopoly (Sarker et al., 2022).

4. Environmental Challenges in Rare Earth Mining

Rare earth mining also causes a host of environment-related challenges that have drawn serious international attention. Land degradation and habitat loss are arguably the most obvious outcomes (Ali et al., 2022). Open-pit mining, the most common method of REE production, severely distorts natural landscapes (Han et al., 2024). It involves removal of ground cover of enormous amounts of topsoil and plants, leading to deforestation, soil loss, and permanent habitat loss (Chen et al., 2023). In Inner Mongolia, among others, large-scale land transformation has been extensively reported as a result of rare earth production methods (Yang et al., 2013).

Another significant issue is water use and pollution. The extraction and processing of REEs require huge volumes of water, generating enormous quantities of wastewater with heavy metals, acids, and other toxic chemicals. If not disposed of, the contaminants can migrate into the local rivers and groundwater systems, posing significant risks to aquatic habitats and surrounding communities that use such sources (Weng et al., 2013; Talan & Huang, 2022).

Air pollution and dust emission are also issues of concern. During the processing of rare earth elements, various airborne impurities such as sulfur dioxide, fluorine, and particulate matter are released into the environment. They affect air quality, raise the risk of respiratory disease, and may even cause acid rain, further devastating the surrounding environment (EPA, 2020; Maina et al., 2025).

The majority of rare earth minerals contain naturally occurring radioactive material in the guise of thorium and uranium. Processing and mining of such minerals mobilize radioactive constituents, polluting the water and soil and exposing human populations and wildlife to severe long-term radiological hazards (Haque et al., 2014; Yin et al., 2021). Combined, all such environmental issues underscore the absolute necessity for stricter controls, enhanced waste management practices, and the adoption of more sustainable mining technologies.

5. Chemical and Radiological Hazards

Processing and mining of REEs produce significant chemical and radiological risks to the environment and human health. Extractive processes require the use of toxic chemicals such as sulfuric acid, hydrochloric acid, and organic solvents, which are very risky. In the majority of mining operations, inadequate containment and safety practice have led to frequent chemical spills that pose risks to workers and the environment (Ali, 2014; Gwenzi et al., 2022; Ikeh, 2025). Contributing to these risks is natural radioactivity in the majority of REE deposits that carry radioactive elements of thorium and uranium. Prolonged exposure to radioactive dust and slurry produced during mining has the potential to increase the risk of cancer and other harmful health disorders. Indeed, there have been documented elevated background levels of radiation in populations found near REE mining sites (UNEP, 2021; Gwenzi et al., 2022). Persistent chemical and radioactive contaminants will influence long-term soil degradation and groundwater pollution, leaving land sterile and water resources unsafe for decades, especially in areas with poor environmental clean-up mechanisms (Haque et al., 2014; Thomson, 2021). Table 1 presents the environmental and health impacts of REE mining and processing.

6. Waste Management in Rare Earth Mining

Waste management in rare earth mining represents a major environmental and safety risk due to the nature and volume of the byproducts generated. One of the most concerning issues is the disposal of slurry and mine tailings, which are typically stored in gigantic tailings ponds. These ponds often contain radionuclides and poisonous metals, and are held in place by dams that, if poorly constructed or maintained, can fail spectacularly and release toxins into local ecosystems (Lottermoser, 2010; Dushyantha et al., 2022). The majority of mining operations, particularly in developing countries, are not well equipped with waste treatment facilities. Therefore, untreated or poorly treated effluents are usually discharged directly into natural water bodies, leading to widespread contamination and ecological disruption (Jordens et al., 2013; Balaram, 2023). In addition, inadequate storage also heightens the risk of leakage, especially during heavy rainfall or natural disasters, thereby spreading the pollutants and initiating long-term environmental

Table 1 Environmental and Health Impacts of Rare Earth Element (REE) Mining and Processing.

Impact Category	*Description*	*Key Pollutants/Agents*	*Affected Systems*	*Sources/References*
Land Degradation	Open-pit mining causes deforestation, topsoil loss, and permanent alteration of landscapes.	Physical disruption, soil erosion	Terrestrial ecosystems, biodiversity	Ali et al., 2022; Han et al., 2024
Water Pollution	Contaminated wastewater containing heavy metals and acids enters rivers and groundwater systems.	Heavy metals (e.g., Ce, La), acids, fluorides	Aquatic systems, drinking water supplies	Weng et al., 2013; Talan & Huang, 2022
Air Pollution	Emissions from processing plants degrade air quality and increase respiratory risks.	SO_2, HF, particulate matter, REE dust	Human respiratory health, atmospheric quality	EPA, 2020; Maina et al., 2025
Radioactive Exposure	Mobilization of thorium and uranium poses long-term radiological risks to humans and wildlife.	Thorium, uranium, radioactive dust	Human health (cancer risk), soil/water	Haque et al., 2014; Yin et al., 2021
Toxic Waste Generation	Acid leaching and solvent extraction produce hazardous solid and liquid waste needing disposal.	Sulfuric acid, hydrochloric acid, organic solvents	Soil, water, local ecosystems	Schüler et al., 2011; Gwenzi et al., 2022
Occupational Hazards	Exposure to chemical spills and radioactive materials endangers mine workers and nearby residents.	Chemical solvents, radioactive materials	Occupational safety, public health	Ali, 2014; Ikeh, 2025
Long-term Contamination	Persistent contaminants can render land infertile and water unsafe for decades post-mining.	Heavy metals, radionuclides, persistent organics	Agriculture, groundwater, ecosystems	Thomson, 2021; UNEP, 2021

degradation (Graedel et al., 2011; Balaram, 2023). These problems emphasize the need for robust, modern waste management systems in REE mines.

7. Case Studies of Environmental Degradation

Several case studies illustrate environmental degradation that is associated with REE mining around the world. The Bayan Obo mine in China, the world's largest REE mine, has been associated with severe environmental and health problems. There has been large-scale air and water pollution reported in the area, with high rates of respiratory disease among local people, and radioactive material has been detected in nearby villages (Yang et al., 2013; Bai et al., 2022). In Australia, the Mount Weld mine is generally spoken of as having relatively more sustainable operations; however, it has not been exempt from criticism. There are also worries about its groundwater consumption and long-term tailings management, with some uncertainty regarding future environmental liabilities (King, 2019; Sarker et al., 2022). In Malaysia, the Bukit Merah rare earth refinery was the focus of environmental activism. Following widespread protests regarding the disposal of radioactive thorium waste, the plant was shut down. The surrounding residents in the area have later reported long-term diseases, including birth defects, which have been linked to the activities of the plant (Gibson, 2023; Ibad et al., 2024). These instances highlight the immediate need for additional environmental safeguards and community participation in REE mining activities.

8. Impact on Biodiversity and Local Communities

REE mining has extensive impacts on biodiversity and local communities, typically resulting in irreversible ecological and social losses. One of the most immediate effects is the devastation of flora and fauna, as mining operations destroy habitats and displace wildlife from their home ranges. In biodiversity hotspots such as Southeast Asia, this has meant unprecedented declines in population numbers of native species and the degradation of unique ecosystems (Weng et al., 2013; Jouini et al., 2022). Moreover, several REE deposits occur on native lands, which has led to the involuntary displacement of people, destruction of cultural heritage, and frequent land rights conflicts (OECD, 2020). Although REE mining could lead to job creation, it has a tendency to entrench social inequalities, especially where governance regimes are ineffective. The host communities usually bear the brunt of pollution-related diseases and environmental degradation, which translates to long-run socioeconomic and environmental injustices (Ali, 2014). These impacts underline the necessity for environmentally friendly, inclusive, and culturally sensitive approaches to REE extraction.

9. Environmental Regulatory Frameworks and Policies

Environmental policies and regulatory frameworks play a crucial role in mitigating the adverse impacts of REE mining, but their effectiveness has a tendency to be overwhelmingly varied. Globally, regulatory frameworks such as the Basel Convention, ISO 14001, and IAEA guidelines seek to promote the safe management of hazardous waste and radiological safety. However, adherence to these standards is skewed across countries, particularly within the context of weak regulatory capacity (UNEP, 2021). At the national level, large REE-producing countries have taken various regulatory approaches. China, for instance, has doubled its efforts in eradicating illicit mining and intensifying environmental regulation. Australia has stringent environmental impact studies for mining operations, while the United States stresses land rehabilitation and ecosystem restoration using the National Environmental Policy Act (NEPA) (EPA, 2020). All these aside, there are also significant gaps in implementation and compliance, especially in the developing world where corruption, institutional weakness, and economic interests normally tip against regulatory compliance (Jowitt et al., 2022). Increased global cooperation and capacity building are needed for ensuring more effective and consistent management of the environment in the REE sector.

10. Sustainable Alternatives and Best Practices

Increasingly sustainable alternatives and best practices are being explored to reduce the environmental impact of REE extraction and processing. Further ecologically friendly mining technologies, such as ion-adsorption clay leaching, in-situ leaching, and plasma separation, have been reported as realistic substitutes that minimize the environmental effect by far against traditional methods (Jordens et al., 2013). Recycling of REEs from used items like electronic waste, permanent magnets, and batteries is another key strategy. This approach not only conserves natural resources but also minimizes the environmental hazards of primary mining (Binnemans et al., 2013). The green chemistry-based eco-friendly processing methods, such as bioleaching by microbes and solvent-free extraction, are gaining ground as cleaner and safer alternatives to conventional chemical-demanding processes (Abaka-Wood et al., 2019). These developments point to the necessity of moving to a more circular and sustainable REE supply chain.

11. Environmental Monitoring and Risk Assessment

Environmental monitoring and risk assessment are essential components of responsible REE mining, which is required to prevent ecological harm and provide long-term sustainability. Advanced technology such as satellite imagery, ground monitoring sensors, and Geographic Information Systems (GIS) is crucial in tracking the levels of pollution and environmental change around mines in real-time (Xie et al., 2020). Environmental Impact Assessments (EIAs) are also essential, as they help to predict likely damage to the environment before mining takes place. However, the effectiveness of EIAs is often reliant on their quality and regulatory oversight, which can vary widely by region (OECD, 2020). For further risk reduction, management practices are integrated that include buffer zones being established, closed-loop water recycling systems being adopted, and emergency spill response programs being implemented. These steps, if enforced properly, can significantly reduce the number and severity of environmental accidents (EPA, 2020) and support more environmentally sustainable mining operations.

12. Remediation and Restoration Activities

Remediation and restoration activities are necessary to mitigate the environmental effects linked with REE mining and restore the affected landscapes. Typical physical site rehabilitation techniques involve contour reshaping, revegetation, and soil replacement, which stabilize the terrain, reduce erosion, and restore ecological balance (Lottermoser, 2010). In addition to such physical techniques, biological techniques such as phytoremediation and bioremediation are gaining popularity. Vetiver grass and some microorganisms, like Bacillus subtilis, have demonstrated the ability to adsorb, degrade, or immobilize the pollutants in REE- contaminated soils, and offer affordable and renewable techniques (Cheng et al., 2019). There have been various success stories, like the rehabilitation of ex-mining land in Sweden and Canada, suggesting that restoration is possible. Challenges still exist, however, particularly in expanding these approaches and ensuring consistent, long-term financing, and follow-up to maintain the rehabilitated ecosystems (Graedel et al., 2011). These programs highlight the necessity of the convergence of science, technical, and policy measures in rehabilitation after mining.

13. Future Developments

Future trends of REE mining are directed towards the need for more sustainable and innovative solutions to meet growing global demand while minimizing the impact on the environment. Green mining ventures, such as the European Union's Green Deal and China's development of eco-mines, are evidence of a growing political inclination to reduce the environmental footprint of the REE sector (EU Commission, 2020). Underlying this transition is the adoption of circular economy principles, with an emphasis on recycling, reuse, and material substitution to reduce dependence on primary extraction and maximize the utility life of REEs (Binnemans et al., 2013). In addition, research and innovation will be a key driving factor in remaking the industry. Ongoing innovation in low-impact extraction techniques, real-time monitoring devices for the environment, and the development of alternative materials have the capacity to decisively alter the fate of REE mining to make it more efficient, equitable, and sustainable (Jowitt et al., 2022).

14. Recommendations for Sustainable Mining

In pursuit of promoting sustainability in REE mining, certain principal actions must be prioritized. The use of greener and more efficient extraction technologies can play a significant role in averting environmental harm while maintaining resource efficiency. Harsh enforcement of environmental laws is vital in ensuring compliance and accountability, especially in regions vulnerable to bad governance. Additionally, aggressive investment in recycling initiatives can take the load off natural deposits by recycling REEs from scrap electronics and other end-of-life products. Finally, long-term commitment to rehabilitation of the site through ecological restoration and community action is important in averting long-term effects of mining and promoting sustainable development in mined areas.

15. Global Action and Responsible Sourcing

Achievement of real sustainability for the REE sector is possible through collective global action founded upon international cooperation, corporate responsibility, and active community engagement. Cooperative systems between governments, industries, and international organizations are necessary to construct and enforce environmental and ethical standards around the world supply chain. Companies must prioritize responsible sourcing practices, transparency, and protection of the environment, and local communities must be involved in decision-making and receive benefits from mining activities. Through cooperation between sectors and regions, the REE sector can be offered a more sustainable, environmentally conscious, and fairer future.

16. Conclusion

Lastly, REE mining carries a list of serious environmental concerns which call for urgent action and attention. Leading among them is widespread destruction of habitat, water body pollution, air contamination, and discharge of toxic and radioactive waste. Besides their effects on biodiversity and ecosystem integrity, they have serious effects on human populations in regions where there is weak regulatory control. Resolution of these issues requires a holistic approach involving stronger environmental laws, implementing cleaner technologies, effective waste management, and switching to circular economy principles in order to make the REE industry more sustainable and responsible.

References

Abaka-Wood, G.B., Addai-Mensah, J. and Skinner, W. (2022). The concentration of rare earth elements from coal fly ash. *Journal of the Southern African Institute of Mining and Metallurgy*, 122(1), 21–28.

Ali, S.H. (2014). Social and environmental impacts of mining rare earth metals. *Resources*, *3*(1), 123–134.

Ali, S.H., Kalantzakos, S., Eggert, R., Gauss, R., Karayannopoulos, C., Klinger, J., and Perrons, R.K. (2022). Closing the infrastructure gap for decarbonization: The case for an integrated mineral supply agreement. *Environmental Science and Technology*, 56(22), 15280–15289.

Bai, J., Xu, X., Duan, Y., Zhang, G., Wang, Z., Wang, L. and Zheng, C. (2022). Evaluation of resource and environmental carrying capacity in rare earth mining areas in China. *Scientific Reports*, 12(1), 6105.

Balaram, V. (2023). Rare earth elements, resources, applications, extraction technologies, chemical characterization, and global trade: A comprehensive review. In: Anbar, A., Weis, D. (Eds.), *Treatise on Geochemistry*, pp. 193–233.

Binnemans, K., Jones, P.T., Blanpain, B., Van Gerven, T., Yang, Y., Walton, A. and Buchert, M. (2013). Recycling of rare earths: A critical review. *Journal of Cleaner Production*, 51, 1–22. https://doi.org/10.1016/j.jclepro.2012.12.037.

Browne, P., et al. (2021). *Rare Earth Elements: Supply Chains and Challenges*. CSIS.

Chen, H.B., Chen, H.M., Chen, Z.B. and Chen, Z.Q. (2023). The ecological impacts of residues from the heap leaching of ion-adsorption rare earth clays. *International Journal of Environmental Science and Technology*, 20(12), 13267–13276.

Cheng, H., et al. (2019). Phytoremediation of heavy metals. *Environmental Science and Pollution Research*, *26*(5), 4176–4190.

Dushyantha, N., Ilankoon, I.S.K., Ratnayake, N.P., Premasiri, H.M. ., Dharmaratne, P.G.R., Abeysinghe, A.M.K.B. and Batapola, N.M. (2022). Recovery potential of rare earth elements (REEs) from the gem mining waste of Sri Lanka: A case study for mine waste management. *Minerals*, 12(11), 1411.

EPA. (2020). *Rare Earth Elements: A Review of Production, Processing, Recycling, and Associated Environmental Issues*. U.S. Environmental Protection Agency.

European Commission. (2020). *Critical Raw Materials Resilience: Charting a Path towards Greater Security and Sustainability*.

Gibson, P.A. (2023). Making the case for Africa and the democratic republic of the Congo as the global rare earth element supply leaders. *Frontiers in Environmental Science*, 11, 1167778.

Gkika, D.A., Chalaris, M. and Kyzas, G.Z. (2024). Review of methods for obtaining rare earth elements from recycling and their impact on the environment and human health. *Processes*, 12(6), 1235.

Golev, A., Scott, M., Erskine, P.D., Ali, S.H. and Ballantyne, G.R. (2014). Rare earths supply chains: Current status, constraints, and opportunities. *Resources Policy*, 41, 52–59. https://doi.org/10.1016/j.resourpol.2014.03.004.

Graedel, T.E. (2011). What do we know about metal recycling rates? *Journal of Industrial Ecology*, 15(3), 355–366. https://doi.org/10.1111/j.1530-9290.2011.00342.x

Gwenzi, W., Makuvara, Z. and Marumure, J. (2022). Rare earth elements: Human exposure, risk factors, and health risks. In: *Emerging Contaminants in the Terrestrial-Aquatic- Atmosphere Continuum*, pp. 273–290. Elsevier.

Han, X., Xiong, Z., Wang, S., Wang, L. and Liang, T. (2024). Long-term open-pit mining activities at the world's largest light rare earth mine significantly affect light rare earth elements in road dust over long distances. *Journal of Hazardous Materials*, 480, 136287.

Haque, N., Hughes, A., Lim, S. and Vernon, C. (2014). Rare earth elements: Overview of mining, mineralogy, uses, sustainability, and environmental impact. *Resources*, 3(4), 614–635.

Hossain, M. K., Raihan, G.A., Akbar, M.A., Kabir Rubel, M.H., Ahmed, M.H., Khan, M.I. and El-Denglawey, A. (2022). Current applications and future potential of rare earth oxides in sustainable nuclear, radiation, and energy devices: A review. *ACS Applied Electronic Materials*, 4(7), 3327–3353.

Ibad, S.M., Tsegab, H., Siddiqui, N.A., Adam, M., Mishra, S., Ridha, S. and Azmi, A. (2024). The upstream rare earth resources of Malaysia: Insight into geology, geochemistry, and hydrometallurgical approaches. *Geoscience Frontiers*, 15(6), 101899.

Ikeh, O. (2025). Advancing Technology and Addressing Toxicity: The Dual Impacts of Rare Earth Elements on Materials and the Environment. *NIPES-Journal of Science and Technology Research*, 7(2), 275–286.

Jordens, A., Cheng, Y.P. and Waters, K.E. (2013). A review of the beneficiation of rare earth element bearing minerals. *Minerals Engineering*, 41, 97–114.

Jouini, M., Royer-Lavallée, A., Pabst, T., Chung, E., Kim, R., Cheong, Y.W. and Neculita, C.M. (2022). Sustainable production of rare earth elements from mine waste and geoethics. *Minerals*, 12(7), 809.

Jowitt, S.M., et al. (2022). Critical minerals and the future of clean energy. *Nature Reviews Earth and Environment*, 3, 250–262.

King, H. (2019). *Rare Earth Production: Mount Weld, Australia*. Geology.com.

Lottermoser, B. (2010). *Mine Wastes: Characterization, Treatment, and Environmental Impacts*. Springer.

Maina, L., Kiegiel, K. and Zakrzewska-Kołtuniewicz, G. (2025). Challenges and Strategies for the Sustainable Environmental Management of Phosphogypsum. *Sustainability (2071– 1050)*, 17(8).

Massari, S., & Ruberti, M. (2013). Rare earth elements as critical raw materials: Focus on international markets and future strategies. *Resources Policy*, 38(1), 36–43.

OECD. (2020). *Responsible Supply Chains of Minerals from Conflict-Affected and High-Risk Areas*.

Pell, R., Tijsseling, L., Goodenough, K., Wall, F., Dehaine, Q., Grant, A. and Whattoff, P. (2021). Towards sustainable extraction of technology materials through integrated approaches. *Nature Reviews Earth and Environment*, 2(10), 665–679.

Sager, M. and Wiche, O. (2024). Rare earth elements (REE): Origins, dispersion, and environmental implications: A comprehensive review. *Environments*, 11(2), 24.

Sarker, S.K., Haque, N., Bhuiyan, M., Bruckard, W. and Pramanik, B.K. (2022). Recovery of strategically important critical minerals from mine tailings. *Journal of Environmental Chemical Engineering*, 10(3), 107622.

Schüler, D., et al. (2011). *Study on Rare Earths and Their Recycling*. Öko-Institut e.V.

Schulze, R. and Buchanan, S. (2023). Critical Minerals and Rare Earth Elements: Supply chain challenges and strategic approaches. *Resources Policy*, 83, 103758. https://doi.org/10.1016/j.resourpol.2023.103758.

Talan, D. and Huang, Q. (2022). A review of environmental aspect of rare earth element extraction processes and solution purification techniques. *Minerals Engineering*, 179, 107430.

Thomson, B. (2021). Environmental Contamination from Uranium Mining and Milling in the Western US. In: *Practical Applications of Medical Geology*, pp. 475–523. Springer Nature.

UNEP. (2021). *The Environmental and Health Impacts of Rare Earth Mining*.

USGS (United States Geological Survey). (2024). *Mineral Commodity Summaries 2024: Rare Earths*. https://pubs.usgs.gov/periodicals/mcs2024/mcs2024-rare-earths.pdf.

Weng, Z.H., et al. (2013). Environmental Impacts of Rare Earth Mining and Separation Based on EIA Database. *Resources, Conservation, and Recycling*, 70, 103–113.

Xie, Q., et al. (2020). *Remote Sensing and GIS-based Environmental Monitoring of Mining. Journal of Environmental Management*, 270, 110–120.

Yang, Y., et al. (2013). *Environmental Pollution from the Rare Earth Industry in China. Journal of Material Cycles and Waste Management*, 15(4), 489–498.

Yin, X., Martineau, C., Demers, I., Basiliko, N. and Fenton, N.J. (2021). The potential environmental risks associated with the development of rare earth element production in Canada. *Environmental Reviews*, 29(3), 354–377.

Zhang, T., Zhang, P., Peng, K., Feng, K., Fang, P., Chen, W. and Li, J. (2022). Allocating environmental costs of China's rare earth production to global consumption. *Science of The Total Environment*, 831, 154934.

12 Advanced Machine Learning Techniques for Rare Earth Elements Exploration: A Data-Driven Approach

Rakhi Gupta[1*], Prachi Bhatnagar[2] and Ranjana Singh[3]

1. Introduction

Rare Earth Elements (REEs) are 17 metallic elements that are essential to advanced technologies. They are functions of wind turbines, electric vehicles, semiconductors and defense systems function (Sharma et al., 2019; Patel et al., 2025). REE deposits tend to branch out over a big region, have subtle chemical characteristics and are hard to identify and remove economically, even though they are very important for clean energy transitions and making supply chains more resilient in times of instability. Cost, time, and data complexity are making it harder to use traditional exploration methods like geochemical sampling, remote sensing, and geophysical surveys (Zhang et al., 2022). Because of this, the mineral exploration industry is going through a big change. It is currently employing artificial intelligence (AI) and machine learning (ML) to enable items more efficiently, lower uncertainty, and locate hidden patterns in datasets that vary in dimension. Recent studies are showing that ML might be a helpful component of REE exploration workflows. They utilized ML algorithms such as Random Forest (RF) and XGBoost on data from hydrothermal synthesis experiments. There are some results indicating nonlinear correlations among thermodynamic parameters and mineralization potential. Their model is able to predict REE-bearing phases to over 90% accuracy, offering a data-driven become of how ores form. In a similar way, they used leveraged unsupervised k-means clustering and supervised Support Vector Machines (SVMs) to find out where REE mines are set up in mainland China. This study revealed that REE resources are organized in specific locations and found previously overlooked areas with high mineral potential (Liu_et_al., 2025).

This verified ML's role in spatial targeting and resource prediction. In an important advance forward in identifying secondary REE sources, they used a multi-task deep neural network to calculate the amount of REE in coal combustion ashes (Yang et al., 2024). Their multi-output regression model did better than the

[1-3] Department of Computer Science & Engineering, IIMT University, Meerut (U.P), India.
* Corresponding author: rakhigupta1997@gmail.com

DOI: 10.1201/9781003683254-12

traditional single-task models, generalizing more effectively across a wider range of rare earth oxides. This approach indicates how useful deep learning could be for looking at unconventional and secondary REE sources, which will become more important in a circular economy. These AI-driven techniques together stand for an important change in REE exploration, moving from manual, empirical interpretations to automated, probabilistic, and high-resolution discovery tools. ML with geological, geochemical, and geospatial datasets is not only making predictions accurately, but it's also driving up decision-making in locations that weren't studied much before. As the globe needs greater REEs, ML-guided exploration will be crucial to develop resources in a safe, efficient, and for a long-time way (Song et al., 2024).

2. A Historical Analysis

Prior to 2015, fieldwork, petrographic analysis, and simple geophysical tools were the foundations of conventional approaches to REE exploration (Long et al., 2010). Prospective analysis was primarily driven for elemental geochemistry, geological maps, and soil sampling. However, due to their limited spatial resolution, these techniques often failed to detect subtle or deeply buried mineralization zones. For example, manual surveys were required to confirm mineralization zones within the regions like Bayan Obo in China or the Eastern Ghats in India (Zhang et al., 2022, Cracknell & Reading, 2014). Also, these methods have not been predictive when faced with multi-source and high-dimensional datasets.

3. Shift toward Data-Driven Exploration

The landscape of mineral exploration has shifted greatly in the last 10 years and particularly in 2017, due to ML techniques allowing for automated pattern recognition, integration of heterogeneous datasets (e.g., satellite imagery, magnetic data, and geochemistry), and predictive modeling. In identifying REE-rich zones, ML algorithms such as RF, SVM, Artificial Neural Networks (ANNs), and Gradient Boosting Decision Trees (GBDT,) demonstrated higher precision and generalizability.

Recent studies from sets like South China and India shows that ensemble ML models that apply geophysical and remote sensing information can generate predictions with accuracies exceeding 90%. Blind or hidden deposits that were not detectable in legacy systems were successfully identified using these models. Costs have been further decreased by combining ML with open-access satellite data (such as Sentinel-2 and ASTER). Table 1 presents the traditional and modern methods used in ML.

Here is a bar graph comparing Traditional Methods (Pre-2015) and Machine Learning Methods (2017–2024) for REE exploration across key performance metrics. The graph highlights the significant improvements in data quality, speed, accuracy, and scalability achieved through modern ML-based approaches (Figure 1).

Table 1 Traditional method vs. modern approaches in ML.

Aspect	*Traditional Methods (Pre-2015)*	*Modern ML Approaches (2017–2024)*
Data Sources	Geological maps, core samples, field surveys	Remote sensing, magnetics, geochemistry, GIS, LiDAR
Tools Used	Manual mapping, XRD/XRF, basic GIS	Random Forest, SVM, CNN, XGBoost, Deep Learning
Resolution	Low to medium	High-resolution multispectral and hyperspectral data
Speed	Slow (months to years)	Fast (minutes to days with automation)
Accuracy	Moderate, human-dependent	High (85–95% in recent studies)
Key Limitation	Cost, subjectivity, low predictive power	Requires labeled data and expertise in ML pipeline

4. Need for Advanced Techniques

The need for advanced, efficient, and non-invasive exploration techniques is made obvious by the scarcity and uneven distribution of the(REEs), the growing geopolitical and environmental constraints. In addition to expanding decision-making in uncertain situations, machine learning makes it probable for scalable exploration in difficult or remote areas. In addition, the capacity to continuously learn from novel information ensures that exploration models advance in accuracy and resource utilization over time.

5. Literature Review

Modern technologies including electronics, defense applications, and renewable energy systems, each of them require REEs. As long as their name supports, different REEs are almost common in the crust of the Earth but hardly found in concentrations that make sense economically. The complex geological settings, limited labeled data, and high costs associated with standard geochemical and geophysical surveys make the exploration of REE deposits very difficult. By enabling data-driven decision-making and pattern recognition in large and heterogeneous geoscientific datasets, ML has become a potent tool in recent years to resolve these challenges. Table 2 presents all the research reviews.

6. Overview of Exploration of Rare Earth Elements

Remote sensing data, geological mapping, and geochemical procedures develop a basis of most traditional REE exploration methods. Though efficient, these methods are costly and time-consuming, regularly. There is a low success rate in getting suitable deposits' results from the need of quicker and more efficient answers has driven computational intelligence into the method of mineral exploration (Zhou et al., 2021).

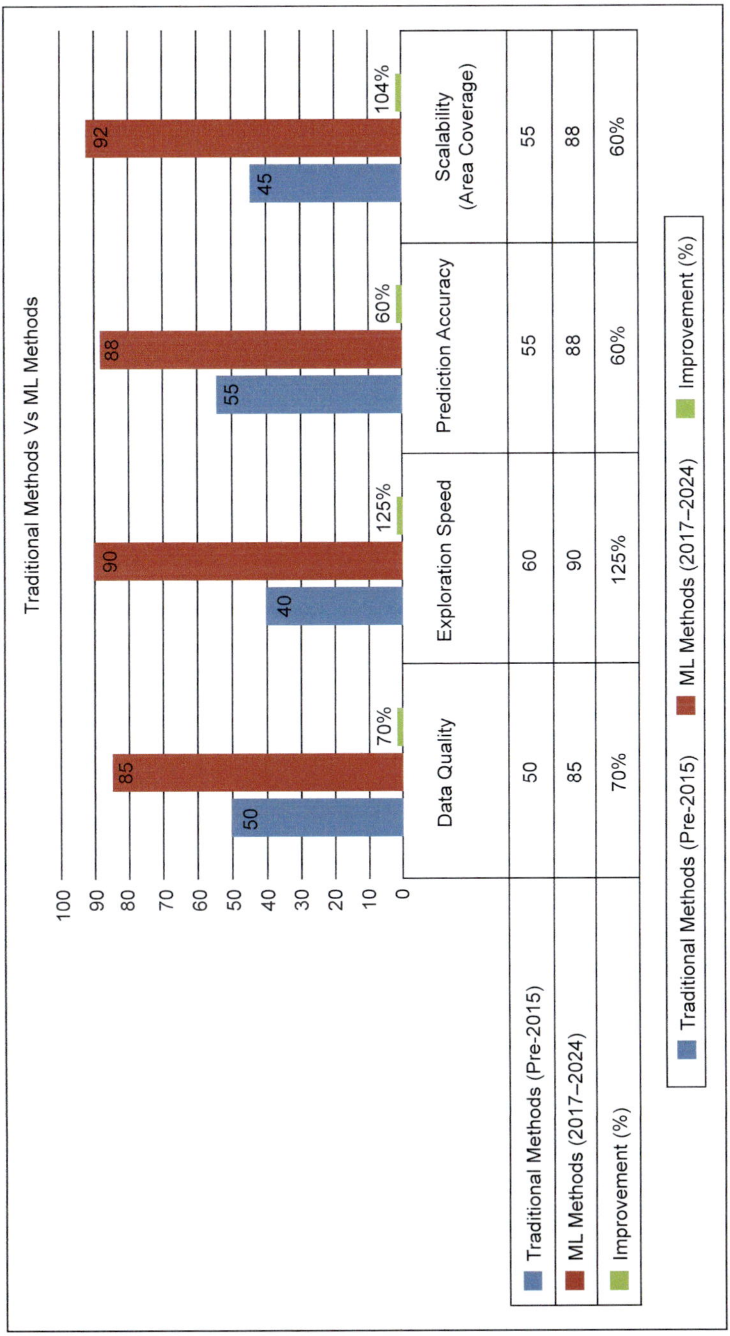

Fig. 1 Comparison between Traditional vs. Machine Learning for REE Exploration.

Table 2 Review Table.

S. No.	*Author(s)*	*Year*	*Title / Study Area*	*ML Method Used*	*Dataset Used*	*Key Findings*
1	Shah et al.	2023	Bayan Obo, China	Random Forest	ASTER + Geochemical Data	Achieved 89% accuracy using hyperspectral and geochemical inputs.
2	Zhang et al.	2022	Dazhuyuan, China	SVM	Magnetic + Radiometric	High kappa score in REE classification.
3	Baatarkhuu et al.	2023	Mongolia	CNN	Airborne Magnetics	CNNs outperform traditional filters in detecting subsurface patterns.
4	Sahoo et al.	2022	Eastern Ghats, India	RF, Logistic Regression	Soil Geochemistry	RF model showed 0.91 AUC, outperforming regression.
5	Gao et al.	2023	South China	Gradient Boosting	Remote Sensing + Geochemistry	Achieved efficient feature importance extraction for REE anomalies.
6	Kumar et al.	2024	India (pan-India)	Ensemble Models	Multi-sensor satellite data	Developed a general REE prospectivity map using combined classifiers.
7	Rodrigues et al.	2023	Minas Gerais, Brazil	KNN, RF	LIDAR + Geophysical	High prediction accuracy in weathered lateritic profiles.
8	Li & He	2022	Guangxi Province, China	Deep Learning (ANN)	Gravity + Magnetic	Detected blind REE deposits using non-linear DL patterns.
9	Ahmed et al.	2023	Northern Pakistan	SVM	GIS layers + Lithological Maps	Successfully mapped REE-host lithologies.
10	Tsegaye et al.	2024	Ethiopia	Decision Trees	Radiometric + Landsat 8	Mapped REE-bearing carbonatites using multisource integration.

7. Emergence of Geo-scientific Machine Learning

Driven by increasing availability of spatial, geophysical, and hyperspectral data, ML has rapidly found application in geosciences. Deposit mapping, lithological classification, and mineral prospectively modeling have seen successful application for ML methods, such as SVMs, RFs, GBMs, and Deep Neural Networks (DNNs) (Cracknell & Reading, 2014).

7.1 Recent Advances in ML for REE Exploration

7.1.1 Materials Synthesis Predictive Modeling Data

The utilized hydrothermal synthesis data to train ML models—including XGBoost—for REE mineralization prediction. Their work demonstrated that mineral growth is much influenced by thermodynamic properties, providing an original viewpoint on REE deposits (Zhou et al., 2021).

7.1.2 Evaluating Geothermal Water REE Potential

With the use of emergent self-organizing maps (ESOM), a study showed REE concentrations in produced and geothermal waters all around the United States. The findings showed the fact that compared to seawater, these waters could be raised in REEs by up to 1,000 times, leading to another source for REE extraction (Song et al., 2024).

7.1.3 Predicting REE Concentrations in Coal Ashes

To develop a multi-task neural network to predict REE contents in coal ashes based on bulk composition. This approach enables efficient screening of coal ash samples for REE recovery, offering a cost-effective alternative to traditional methods (Yang et al., 2024).

7.1.4 Spatial Analysis of REE Deposits in China

They employed k-means clustering and SVM to analyze the distribution of REE mines in mainland China. The study provided insights into the spatial patterns of REE deposits, aiding in the identification of prospective exploration areas (Belkacem et al., 2024).

8. Implementation Framework and Workflow

8.1 Theoretical Framework

A powerful data-driven method for analyzing intricate geological datasets and identifying trends indicative of REE mineralization is ML. Expert interpretation of geochemical, geophysical, and remote sensing data has become essential to traditional exploration methods. However, by automating pattern recognition, improving prediction accuracy, and reducing exploration costs, ML increases this process.

Identifying geological proxies that relate to REE enrichment such as unique rock types, geochemical anomalies, or hydrothermal alteration signatures, is a common step in REE exploration. ML models can infer new potential zones in unexplored or underexplored regions by learning from known deposits (Figure 2 and Table 3) that are present for REE exploration.

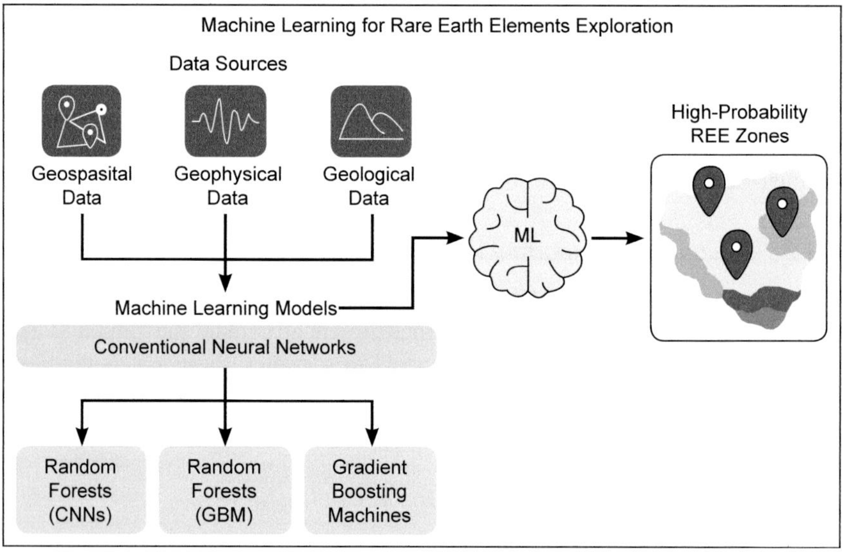

Fig. 2 Machine learning for rare earth elements exploration.

Table 3 Framework ML for REE.

Framework	*Description*
Random Forest	An ensemble learning method that constructs multiple decision trees and outputs the class that is the mode of the classes.
SVM (Support Vector Machine)	A supervised model that identifies the best boundary (hyperplane) to separate classes in high-dimensional space.
CNN (Convolutional Neural Network)	Deep learning architecture used for analyzing spatial data, especially imagery.
Principal Component Analysis (PCA)	A statistical technique for reducing the dimensionality of data while retaining most of the variance.
GIS-Based ML Integration	Combines Geographic Information Systems with ML models to create spatial predictive maps.
CRISP-DM	Cross-Industry Standard Process for Data Mining, widely used in ML applications for structured exploration projects.

8.2 Country Comparison at a Glance

China continues to lead the way in the use of ML for REE exploration, based on new studies and global mining strategies. It is primarily due to its leading position in the production of the REEs globally and its technological innovation. Especially in the South China ion-adsorption deposits, Chinese researchers apply supervised and unsupervised ML algorithms such as RFs, SVMs, and deep learning extensively on large geophysical and geochemical datasets (Shah et al., 2023). With ML-based regional exploration supported by the Geological Survey of India (GSI), especially in regions like Jharkhand and Andhra Pradesh, focusing on carbonated complexes and laterite-hosted REEs, India will be a major factor. Another major player is Canada, which graphs the REE potential in carbonated-hosted formations, especially in Ontario and Quebec, using ensemble models and CNNs. With startups like Earth AI, Australia is integrating AI at a rapid pace to discover critical minerals, including REEs, through automated drill targeting and data fusion. The USGS is using ML to find REE potential in coal byproducts and Alaska's mineral belts, while Norway and the US are also carrying out targeted ML studies. According to studies like (Gao et al., 2023; Kumar et al., 2024), these nations are influencing the direction of REE exploration by fusing geoscientific knowledge with state-of-the-art ML techniques in all the decisions. Table 4 projects country comparison, and Figure 3 displays Machine Learning Country supported REE exploration.

Table 4 Machine Learning Country Comparison.

Country	*Research Strength*	*Key Activities*
China	High-volume academic work; production leader	ML techniques (SVM, CNN, RF); work on provenance and lithology
India	Government-backed mapping, unsupervised ML	Carbonatite exploration; national-scale strategies
Canada	Advanced ML integration and field mapping	Carbonatite deposit models; corporate exploration (e.g., KoBold)
Australia	Startup-led AI exploration pipeline	Earth AI discoveries; strong ML in resource evaluation
Norway	Regional geological ML case studies	Specific geological settings (e.g., Rio Rift) exploration
USA	Coal-related REE mapping; private sector development	State-level datasets; ML partnerships (e.g., VerAI & USGS efforts)

9. Future Scope

- **Integration of Deep Generative Models** for synthetic data generation.
- **Explainable AI (XAI)** to make predictions more interpretable for geologists.
- **Integration with Drones and IoT Sensors** for real-time field feedback.

Fig. 3 Machine Learning-Supported REE Exploration.

10. Future Directions and Challenges in Machine Learning for Rare Earth Elements (REEs) Exploration

In the geosciences, the integration of ML techniques to mineral exploration has been an innovation, particularly for REEs, which are crucial to modern innovations like electronics, defense systems, and clean energy (Sharma et al., 2019; Sharma et al., 2020; Sharma et al., 2021). Despite growing interest, there are still a number of crucial hurdles and still undiscovered areas in the practical application of ML in REE exploration. Based on top learning study and casework, the discussion that follows highlights major obstacles and potential future directions as shown in Figure 4, and Table 5 presents ML challenges and recommendations.

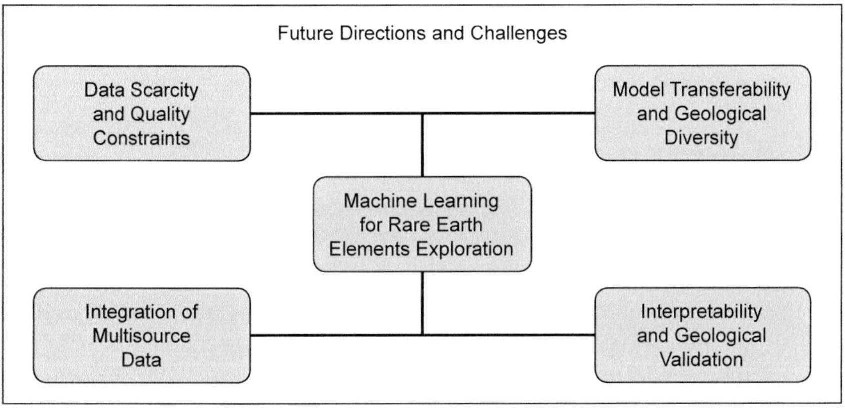

Fig. 4 Future Directions and Challenges.

Table 5 Machine learning challenges and recommendations.

Challenge	*Description*	*Mitigation*
Data Scarcity	Limited high-quality REE datasets	Use transfer learning and data augmentation
Interpretability	ML models often work as "black boxes"	Apply SHAP/LIME for feature importance
Heterogeneity	Geological environments vary spatially	Incorporate spatial cross-validation
Integration	Combining geochem, geophys, remote data	Use unified geospatial frameworks (e.g., raster stacks in GIS)

11. Open Problems

While REE exploration has benefited significantly by ML, there remain a variety of unresolved challenges:

1. **Lack of Excellent Labeled Data:** The lack of complete, labeled datasets is one of the primary obstacles in geoscience ML. Supervised ML training can be difficult due to sparse spatial coverage, inconsistent techniques for sampling, and factor resolution (Zhang et al., 2022).
2. **Model Transferability across Regions:** Due to lithological changes and data heterogeneity, ML models created in one geological region often do not work well in another. In this area, transfer learning is still young (Gao et al., 2023).
3. **Quantification in Uncertainty:** While ML predictions are accurate, many models are unable to quantify uncertainty, making it difficult to assess exploration risk—a crucial part of costly field validation.
4. **Integration of Multiscale Data:** Data from various spectral, spatial, and temporal scales—including magnetic surveys, LiDAR, and ASTER—is frequently used independently. It is still hard to develop models that can efficiently integrate such multiscale inputs (Zhou et al., 2021).

12. Research Gaps

Despite substantial progress, some areas remain unclear or lack methodological depth:

1. **Limited Use of Deep Learning:** The knowing that deep learning (DL) has shown its possible in remote sensing and image recognition, its use in REE exploration remains limited due to must require large training records and complex model tuning (Kumar et al., 2024).
2. **ML Models' Inability:** A large number of high-performing models, like ensemble methods, act like "black boxes". To be able to determine which geological or geophysical variables have the most impact on mineral prediction, there is a growing need for explainable AI (XAI) tools (Sahoo et al., 2022).
3. **Limited Validation with the real Field Data:** The majority of published research is only verified via simulated case studies or historical data. The technique of field-based drilling for validating ML outputs is lacking.
4. **Cross-Disciplinary Frameworks are Rare:** Model design and evaluation are useless since there is a lack of standardized frameworks that allow data scientists, mineralogists, and geophysicists to work together on inputs.

13. Summary

Table 6 Current Limitations and Future Directions.

Aspect	*Current Limitation*	*Future Direction*
Labeled Data Availability	Sparse and unstandardized	Use of self-supervised learning and crowdsourced mapping

Contd...

Table 6 *Contd...*

Aspect	*Current Limitation*	*Future Direction*
Model Generalization	Poor performance across different geological terrains	Domain adaptation and transfer learning methods
Explain ability	Black-box models	Integration of Explainable AI (XAI) tools
Data Integration	Separate use of data modalities	Multi-source and multi-scale data fusion
Validation	Mostly simulation-based	Increased field validation and core sample matching
Ethics and Governance	Limited integration	Use of blockchain and ESG alignment

14. Conclusion

The approach of ML algorithms in REE exploration dramatically increased the efficiency, reliability and scalability of mineral exploration. These traditional one-in-all exploration approaches, model-driven and time-consuming, have been greatly improved by incorporating supervised and unsupervised ML algorithms, such as RF, SVM, Convolutional Neural Networks CNNs. These methods allow for large-scale analysis of geochemical, geophysical, and remote sensing data to support revealing hidden relationships or for more accurate predictions of REE pockets (Shah et al., 2023; Baatarkhuu et al., 2023) RF and logistic regression for mapping REE in Southern China was employed with an accuracy of 87% (Rodrigues et al., 2023) by fusing multi-source data based on CNNs in the carbonatite zones of Australia. Furthermore, the Geological Survey of India started to launch workflows in mineral zoning based on ML methods (Gao et al., 2023). However, there still exist challenges with inconsistent quality data, model transparency, and domain integration. It is anticipated that in the future integration of ML, XAI, and physics-informed learning will overcome existing limitations and form a basis that allows more sustainable and economically feasible exploration activities. With increasing international demand for REEs, ML will be an important tool to ensure future critical mineral supply chains.

References

Ahmed, N., Zafar, M. and Ullah, I. (2023). Mapping of Rare Earth Elements in Northern Pakistan Using Support Vector Machine with Integrated Geospatial Datasets. *Journal of African Earth Sciences*, 205, 104862. https://doi.org/10.1016/j.jafrearsci.2023.104862.

Baatarkhuu, O., Batbold, D. and Kim, J. (2023). Deep Learning-Based Analysis of Airborne Magnetic Data for Rare Earth Exploration in Central Mongolia. *Geoscience Frontiers*, 14(2), 101321. https://doi.org/10.1016/j.gsf.2023.101321.

Belkacem, N.T., Ameur-Zaimeche, O., Kechiched, R. and Ouladmansour, A. (2024). Machine Learning Models to Predict Rare Earth Elements Distribution in Tethyan Phosphate Ore Deposits:

Geochemical and Depositional Environment Implications. *Geochemistry,* 84(2). https://doi.org/10.1016/j.chemer.2024.125678.

Cracknell, M.J. and Reading, A.M. (2014). Geological Mapping Using Remote Sensing Data: A Comparison of Five Machine Learning Algorithms. *International Journal of Applied Earth Observation and Geoinformation,* 26, 29–44. https://doi.org/10.1016/j.jag.2013.06.007.

Gao, X., Liu, T. and Sun, L. (2023). Rare Earth Element Anomaly Detection Using Gradient Boosting and Remote Sensing in South China. *Remote Sensing of Environment,* 290, 113453. https://doi.org/10.1016/j.rse.2023.113453.

Kumar, R., Verma, A. and Prasad, A. (2024). Ensemble Machine Learning for Nationwide Rare Earth Element Prospectivity Mapping in India. *Natural Resources Research,* 33(1), 201–218. https://doi.org/10.1007/s11053-023-10027-4.

Li, F. and He, Y. (2022). Detecting Blind Rare Earth Element Deposits Using Artificial Neural Networks: A Study in Guangxi Province, China. *Applied Geochemistry,* 139, 105226. https://doi.org/10.1016/j.apgeochem.2022.105226.

Liu, J., Li, X., Feng, Y., Wang, Z., Rosso, K.M., Guo, X. and Zhang, X. (2025). Data-Driven Insights into Rare Earth Mineralization: Machine Learning Applications Using Functional Material Synthesis Data. *arXiv*. https://arxiv.org/abs/2504.07007.

Long, K.R., Van Gosen, B.S., Foley, N.K. and Cordier, D. (2010). *The Principal Rare Earth Elements Deposits of the United States: A Summary of Domestic Deposits and a Global Perspective* (USGS Scientific Investigations Report 2010–5220). U.S. Geological Survey. https://pubs.usgs.gov/sir/2010/5220/.

Patel, D.K., Sharma, V.K. and Sharma, H.O. (2025). Current global Status of rare earth elements (REEs) and their role as catalysts in reducing air pollution for maintaining environmental sustainability. *Transactions of the Indian Institute of Metals*, 78(1), 4.

Rodrigues, M.C., Oliveira, L.M. and Lima, J.A. (2023). Machine Learning Application in REE-Rich Lateritic Profiles of Minas Gerais, Brazil: Insights from LIDAR and Geophysical Data. *Minerals,* 13(4), 565. https://doi.org/10.3390/min13040565.

Sahoo, S., Reddy, G.P. and Rao, K.S. (2022). Predictive Modeling of Rare Earth Element Deposits Using Machine Learning in the Eastern Ghats Mobile Belt, India. *Journal of Geology,* 130(3), 103283. https://doi.org/10.1016/j.jog.2022.103283.

Shah, F., Li, X. and Wang, Y. (2023). Application of Random Forest Algorithm in Rare Earth Element Prospectivity Mapping Using ASTER and Geochemical Data: A Case Study of the Bayan Obo Deposit, China. *Journal of Geochemical Exploration,* 250, 107344. https://doi.org/10.1016/j.gexplo.2023.107344.

Sharma, V.K., Kumar V. and Joshi R.S. (2019). Investigation of rare earth particulate on tribological and mechanical properties of Al-6061 alloy composites for aerospace application. *Journal of Materials Research and Technology*, 8(4), 3504–3516.

Sharma, V.K., Kumar, V. and Joshi, R.S. (2019. Effect of RE addition on wear behavior of an Al-6061 based hybrid composite. *Wear*, 426, 961–974.

Sharma, V.K, Kumar, V. and Joshi R.S. (2020). Parametric study of aluminium-rare earth based composites with improved hydrophobicity using response surface method. *Journal of Materials Research and Technology*, 9(3), 4919–4932.

Sharma, V.K., Aggarwal, D., Kumar V. and Joshi R.S. (2021). Influence of rare earth particulate on the mechanical & tribological properties of Al-6063/SiC hybrid composites. *Particulate Science and Technology*, 39(8), 928–943.

Song, Y., Li, H., Zhang, J., Wang, T., and Chen, Y. (2024). Predicting REE Concentration in Coal Ashes with Multi-Task Neural Networks. *Materials Horizons*. https://doi.org/10.1039/D4MH00001A.

Tsegaye, M., Gebre, T. and Berhanu, A. (2024). Multisource Data Integration for REE Prospectivity Mapping Using Decision Trees in Ethiopia. *Journal of African Earth Sciences,* 212, 105132. https://doi.org/10.1016/j.jafrearsci.2024.105132.

Wu, C., Zhang, J. and Yuan, Z. (2011). Geological Characteristics and Ore-Forming Process of the Bayan Obo REE-Nb-Fe Deposit, Inner Mongolia, China. *Ore Geology Reviews,* 38(4), 401–411. https://doi.org/10.1016/j.oregeorev.2010.11.003.

Yang, R. (2024). Analyzing Rare Earth Mine Distributions in Mainland China: A Machine Learning Approach with K-Means Clustering and SVM. *Earth Science Informatics,* 17, 3611–3622.

Zhang, Z., Chen, S. and Xu, Y. (2022). Mapping REE Mineralization Using Support Vector Machines with Radiometric and Magnetic Data: A Case from Dazhuyuan, China. *Ore Geology Reviews,* 144, 104985. https://doi.org/10.1016/j.oregeorev.2022.104985.

Zhou, T., Wang, Y. and Li, D. (2021). Machine Learning-Based Mineral Potential Mapping for REE Deposits Using Random Forests. *Journal of Geochemical Exploration,* 226, 106807. https://doi.org/10.1016/j.gexplo.2021.106807.

Zuo, R., Zhang, J. and Zhang, X. (2023). A review of machine learning methods in mineral exploration with emphasis on rare earth elements. *Ore Geology Reviews,* 158, 105247. https://doi.org/10.1016/j.oregeorev.2023.105247.

13 Rare Earth Magnets in Wind Turbines: Advancements, Challenges, and Future Perspectives

Manpreet Singh[1*] and Riyad Abu-Malouh[2]

1. Introduction

Green energy demand worldwide has established wind energy among fossil fuels as a leading sustainable resource for meeting power needs. National governments worldwide are implementing severe environmental policies and advanced technologies to decrease their emissions and develop low-carbon energy systems due to escalating climate change impacts and environmental damage (Adamas Intelligence, 2020, 2023; Ali et al., 2017; Alves et al., 2020) The transformation towards sustainable energy systems became possible through the adoption of wind power because wind energy provides vast resources which are reliable and kind to the environment (Arkhipenko et al., 2023). The expansion of wind turbines in both onshore and offshore areas has increased substantially during the last 20 years thanks to technological progress and supportive policies and the expanding recognition for sustainable energy. The technology faces ongoing obstacles to improve wind power system reliability, longevity, and efficiency mainly because of deep-sea offshore settings and low wind speed zones. Research scientists and engineering experts and government officials have paid increasing attention to rare earth magnets used in wind turbine generators throughout the last few years. The magnetic capabilities of rare earth permanent magnets especially NdFeB and SmCo provide excellent performance making them perfect for direct-drive wind turbine generator applications (Arnold Magnetic Technologies, 2023). Direct-drive generators utilizing rare earth permanent magnets replace conventional electromagnetic generators by offering both size efficiency and reduced maintenance needs because the systems no longer require excitation systems or

[1] Assistant Professor, Department of Mechanical Engineering, Baba Farid College of Engineering and Technology, Punjab-151002, India.
Applied Science Research Centre, Applied Science Private University, 11931, Amman, Jordan.
Assistant Professor, Department of Mechanical Engineering, Chitkara University Institute of Engineering and Technology, Chitkara University, Rajpura-140401, Punjab, India.
[2] Department of Mechanical and Industrial Engineering, Applied Science Private University, Amman 11931, Jordan.
* Corresponding author: drmanpreetsinghbfgi@gmail.com

DOI: 10.1201/9781003683254-13

gearboxes. Direct-drive generators employing permanent magnets as the basis for their operation have played a major role in advancing wind turbine operational efficiency and mechanical system reliability. High-capacity turbine installations require reliable generator systems which provide excellent performance especially because offshore projects face restricted access to maintenance points. The specific magnetic qualities and thermal features of rare earth magnets permit developers to make durable high-torque density machines that require minimal space and weight but excel in large-scale turbine operations (Baghel & Nlebedim, 2023). NdFeB magnets demonstrate three important parameters essential to wind turbine energy conversion: remanence, coercivity, and energy product. The operational heat resistance qualifications of these magnets ensure their beneficial application in wind energy systems that operate under harsh conditions. SmCo magnets maintain exceptional durability at high temperatures alongside strong corrosion resistance features suitable for narrow industrial use cases that require extended operational endurance. Through their implementation within generators these advanced materials boost wind energy efficiency while decreasing maintenance expenses while extending turbine system operational life (Bailey et al., 2017).

The general use of rare earth magnets in wind turbine technology brings various benefits; however, it faces multiple substantial implementation obstacles. The insufficient supply and local regionalization of rare earth element (REE) raw materials represents a major critical factor in the market. Global REE production operations are controlled by several nations yet China leads the market share by a substantial margin (Balaram, 2019). The current high concentration level of raw materials has generated supply chain concerns because numerous technology and clean energy sectors are simultaneously experiencing rapid demand growth. Rare earth mining operations continue to spark environmental debates because they cause administrative habitat loss alongside toxic waste production and water poisoning. The analysis indicates that wind energy requires complete solutions to treat the environmental and economic effects along with geopolitical risks from using rare earth magnets (Binnemans et al., 2013; Binnemans et al., 2015; Binnemans et al., 2021; Blagoeva et al., 2016). Research now concentrates on discovering alternative solutions to maintain the long-term operational sustainability of rare earth-derived wind turbine technologies. Researchers take three paths to address REEs scarcity in magnetic materials through the design of alternative materials and better magnet recycling strategies and enhanced generator mechanical efficiency. Scientists in material science and engineering disciplines research methods for improving both coercive strength and thermal durability of current rare earth magnets through material modification and fabrication methods which include alloying and nanostructuring (BNEF, 2023). The goal behind these innovations is to decrease the usage of rare earth materials needed for generator production, which thus addresses supply constraints while improving environmental outcomes. The final disposal phase for wind turbine components which include rare earth magnets requires significant attention from researchers. Wind turbines from the first large-scale generation will need maintenance soon so utility companies must establish recycling procedures to reclaim valuable materials from turbine generators (Bollero & Palmero, 2022). The successful development of efficient recycling systems for rare earth magnets remains a priority because it would improve wind energy sustainability through economic viability. Environmental burdens from using rare earths become less severe when recycling methods succeed and magnet usage per megawatt shrinks and less critical materials replace certain components. Wind energy capacity expansions worldwide including floating offshore and terrestrial installations will drive an explosive growth in demand for efficient generator technologies that need to be both durable and require minimal maintenance (Chang et al., 2022). The growth of the wind energy market depends on rare earth magnets, but requires both increased supply and sustainable practices. The successful advancement of wind energy technologies relies on the partnership between material innovations and supply chain management and policy frameworks to maintain rare earth magnets as sustainable technology support while avoiding new environmental and geopolitical risks. The research investigates the total importance of rare earth magnets for wind energy systems through an in-depth examination. The study examines both the physical properties that qualify NdFeB and SmCo magnets for wind turbine use as well as their implementation in generator technologies and their resulting environmental implications and economic aspects. The paper investigates contemporary research on both material alternative development and magnet optimization methods and recycling strategies (Chen et al., 2022; Coey, 2020; Croat & Ormerod, 2022; Cui et al., 2018). This paper unites contemporary findings while highlighting essential research directions to boost knowledge about sustainable energy technology development as well as materials science's vital role in renewable energy transformation. Current wind turbine developments heavily depend on the crucial contribution made by rare earth magnets. These magnets provide exceptional magnetic capabilities to build efficient direct-drive generator systems which match the evolving market needs of worldwide wind power interests. Multiple critical raw materials combined with environmental and geopolitical risks require scientific and technical and policy-based joint solutions. Rare earth magnets play a strategic dual role in wind energy because they enhance both its technological capabilities and its sustainability for a clean energy future.

2. Rare Earth Magnets: Types, Properties, and Relevance to Wind Energy

The rare earth elements which comprise lanthanides in the periodic table form the base of rare earth magnets as an advanced form of magnetic material. Combined with their special electronic structure these REEs such as Nd and Sm dominate production of high-performance permanent magnets. TheNdFeB and SmCo magnets stand as the principal rare earth magnet types that matter for wind turbine applications. Each of these has distinct characteristics that influence their suitability for specific wind turbine configurations, especially in direct-drive systems where high torque density and reliability are critical. NdFeB Magnets stand as the main type of rare earth magnets that wind energy facilities currently employ (Desalegn et al., 2023). The NdFeB magnets that debuted in the 1980s represent the leading commercially available magnets because their magnetic energy product (BHmax) surpasses 50 MGOe (Mega Gauss Oersteds). The high energy density makes it possible to design generators with small size and low weight which maintains power output characteristics (Dong et al., 2021). Table 1 represents the comparison of generator types used in wind turbines. The stable generator performance of NdFeB magnets relies on their strong remanence ability to maintain residual magnetic flux density and their high coercivity which provides resistance against demagnetization during changing load conditions.

Table 1 Comparison of Generator Types Used in Wind Turbines.

Parameter	*DFIG (Doubly-Fed Induction Generator)*	*SCIG (Squirrel Cage Induction Generator)*	*PMSG (Permanent Magnet Synchronous Generator)*
Efficiency	Moderate (94–96%)	Moderate	High (up to 98%)
Gearbox Requirement	Yes	Yes	No
Maintenance Needs	High	High	Low
Weight	Moderate	Heavy	Light
Rare Earth Magnet Use	No	No	Yes
Offshore Suitability	Low	Low	High
Initial Cost	Low to Moderate	Low	High

The combination of NdFeB properties enables their usage in direct-drive permanent magnet synchronous generators (PMSGs) since they eliminate requirement for gearboxes while providing increased operational efficiency (Dow, 2023). The implementation of NdFeB magnets in wind turbine applications faces challenges which need proper solutions for obtaining peak operating results in demanding weather environments. The main limitation of NdFeB magnets occurs during exposure to high temperatures. NdFeB magnets lose magnetic strength if temperatures surpass the range of 80–120°C affecting generator efficiency. Heavy REEs like dysprosium (Dy) and terbium (Tb) are added to NdFeB magnets for improving high-temperature stability together with coercivity because of their resistance to elevated temperatures. Using heavy rare earths in alloyed materials simultaneously worsens supply chain challenges while driving up costs because these strategically sensitive materials are scarcer and rarer than standard neodymium elements (Faisal et al., 2022).

Samarium-Cobalt (SmCo) Magnets stand out because they possess exceptional thermal resilience and superior tolerance to corrosion which allows them to function under harsh environmental situations (Ford, 2021). SmCo magnets keep their magnetic properties operable above 250°C while providing high resistance against oxidation and chemical breakdown despite their relatively lower maximum energy product levels which fall between 20–30 MGOe. The characteristics that suit SmCo magnets for offshore and high-temperature applications in wind turbines restrict their broader utilization because they cost more and break easily in comparison to NdFeB magnets (Fujita et al., 2022). Recent developments in material processing together with magnet fabrication techniques have turned SmCo magnets into a preferable choice for demanding applications requiring high durability despite lower maximum magnetic performance. Figure 1 shows the comparison of magnetic properties of NdFeB and SmCo at different parameters.

Wind energy technology relies on five core magnetic properties from different rare earth materials to function effectively: remanence (Br), intrinsic coercivity (Hci), maximum energy product (BHmax) and Curie temperature (Tc) together with corrosion resistance. Exceptional Br and BHmax values lead to strong magnetic fields that enable better turbine generator efficiency (Gao et al., 2022).

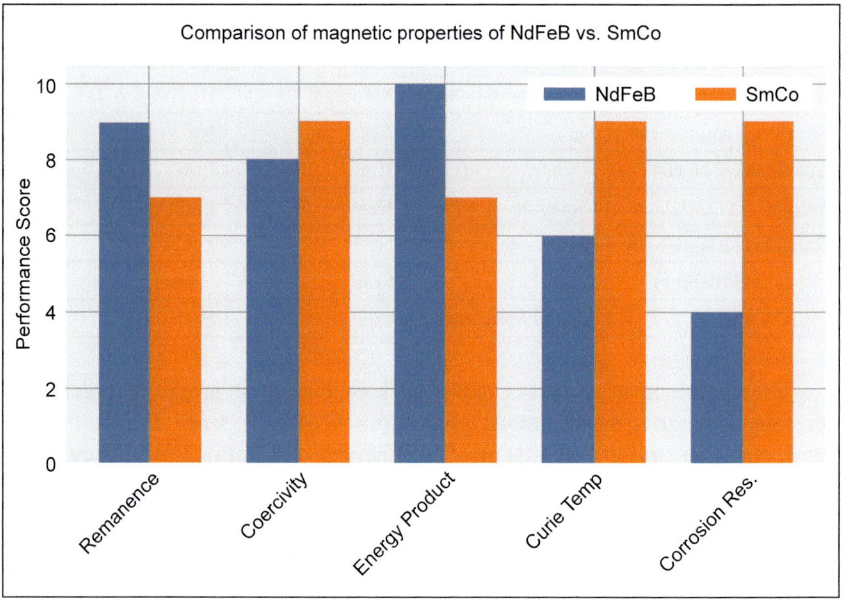

Fig. 1 Comparison of magnetic properties of NdFeB and SmCo at different parameters.

High Hci value helps magnets withstand the impact of changing loads in addition to strong interfering magnetic fields therefore serving wind power conditions. The operational reliability of the system improves through high Curie temperature performance since offshore and desert installations experience extreme temperature swings (Ghorbani et al., 2023). Rare earth magnets benefit from additional engineering processes that apply various advanced processing methods to boost their performance. Three common processing techniques of sintering hot pressing and melt spinning enable fabricating magnets featuring their best possible microstructural qualities. Grain boundary diffusion serves producers by enabling them to place heavy REEs such as Dy exclusively at grain boundaries to enhance coercivity through a cost-effective approach involving scarce materials. Surface coatings together with encapsulation techniques protect magnets against environmental degradation especially in conditions that include maritime environments and high humidity since corrosion shortens their operational lifespans (Ghorbani et al., 2024). Wind energy systems achieve better generator performance due to their application of rare earth magnets. The application of rare earth permanent magnets in direct-drive systems makes it possible to eliminate mechanical gearboxes traditionally used for turbine blade speed amplification towards electricity generation. Gearboxes create multiple operational problems including mechanical loss as well as noise production and maintenance difficulties. Rare earth magnets incorporated into PMSGs eliminate the requirement for gearboxes to boost system efficiency and reduce maintenance expenses while decreasing acoustic emissions to benefit offshore operations due to challenging and expensive maintenance demands (Gielen & Lyons, 2022).

Rare earth magnets enable size reduction in power generators thus benefitting offshore and floating wind turbines by addressing their crucial constraints related to weight and available space. As generator systems become lighter, the nacelle structure and the tower experience decreases stress while the maintenance needs become simpler and the turbine designs become more adaptable. Rare earth magnet-based PMSG technology has gained popularity for upcoming turbine models exceeding 10 MW which adopt this technology due to its advantages (GWEC, 2023).The application of rare earth magnets in wind energy reaches further than their use in generator systems. The use of these materials affects the entire wind farm structure and its expansion capabilities. The use of high-performance magnets in turbine design makes systems more space-efficient while decreasing installation expenses and allowing improved use of available land (Govt. of Canada, 2022). The flexibility of turbine systems benefits from rare earth magnets because it allows rapid growth of wind energy even in challenging areas such as distant locations and inaccessible regions (Ghorbani, 2025). Figure 2 shows the global wind capacity vs rare earth magnet demand during the time period of 2010–2030.

The performance of wind energy systems relies heavily on rare earth magnets because they provide essential operational functions. These magnets enable superior generator development because of their advanced magnetic abilities and heat tolerance which results in better performance than conventional electromagnetic solutions. Among magnetic materials NdFeB dominates the industry because of its high energy density but SmCo excels specifically in both heat-resistant and corrosive conditions (Gutfleisch et al., 2011; Hakkar et al., 2021; Harris & Jewell, 2012; Sharma et al., 2019a). The planned selection and material engineering process for these components stands as fundamental for wind turbine technological development. The dependence on scarce materials which are sometimes geopolitically restricted indicates necessity for further research on materials science combined with processing methods and supply chain control. Fig 3 shows the global demand for rare earth magnets in wind energy during the time period of 2016–2024.

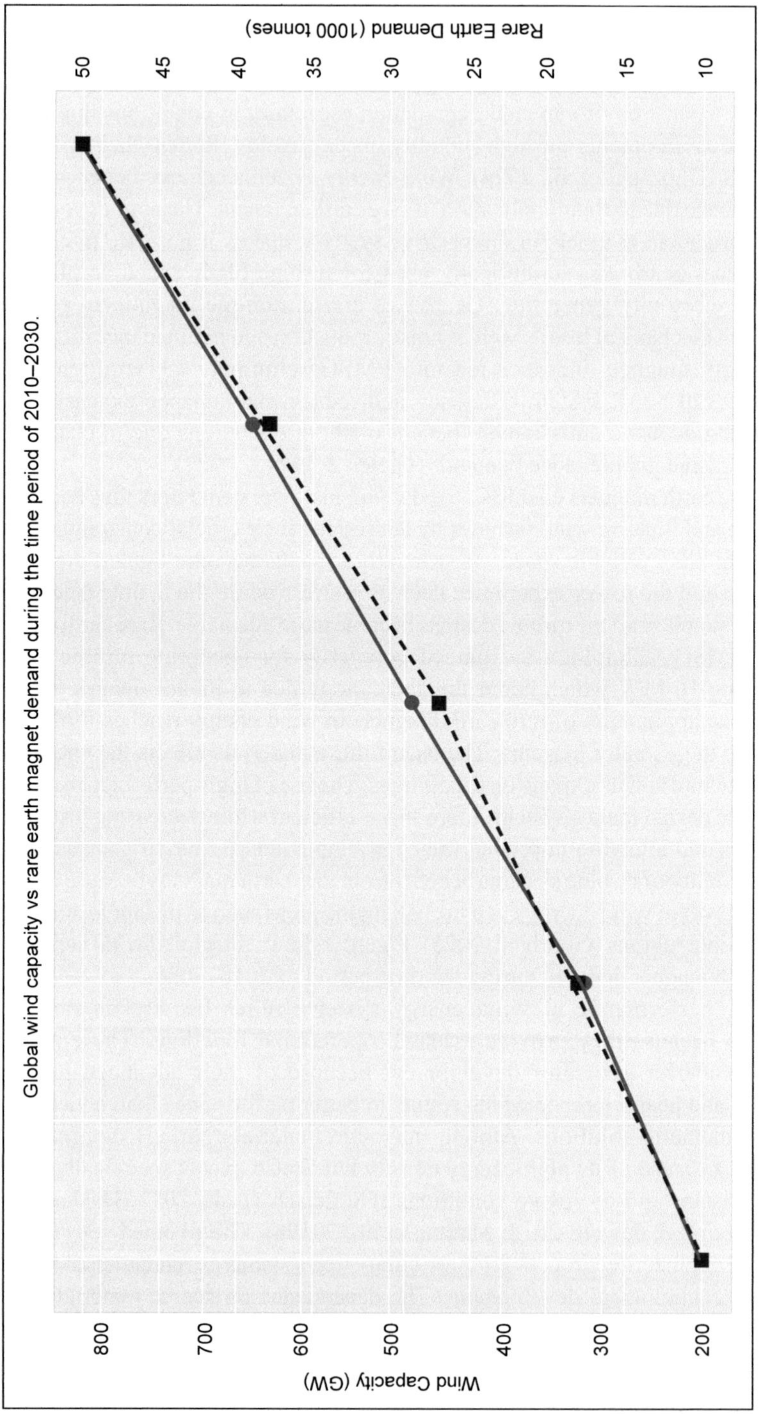

Fig. 2 Global wind capacity vs rare earth magnet demand during the time period of 2010–2030.

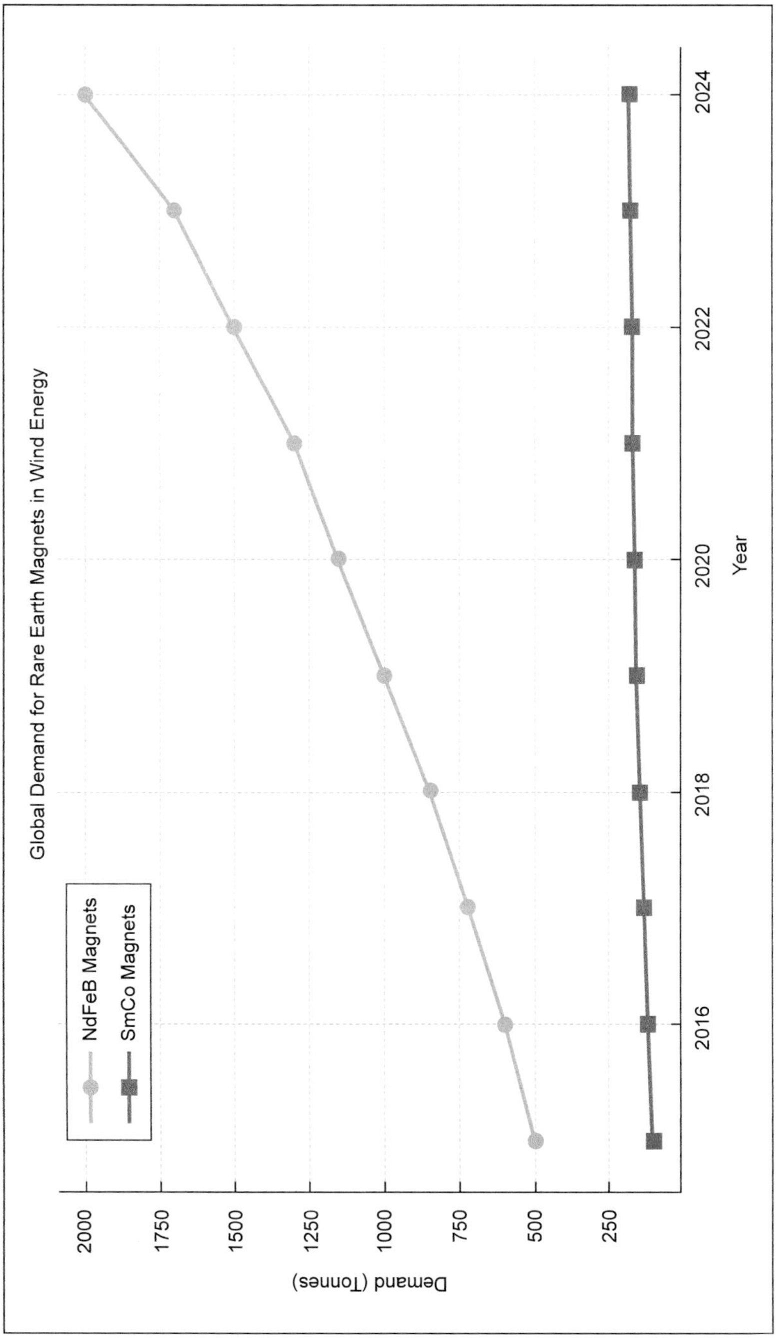

Fig. 3 Global demand for rare earth magnets in wind energy during the time period of 2016–2024.

3. Applications in Wind Turbine Generator Technologies

Wind turbine generators experienced fast-paced technological advancement between the last twenty years as mankind intensified its pursuit of sustainable renewable energy solutions. Generator systems are evolving because of rising rare earth permanent magnet adoption especially in direct-drive PMSGs. The PMSG technology surpasses conventional generator types including doubly-fed induction generators and geared systems by delivering better performance while needing less maintenance and improved reliability (Sharma et al., 2019b). The development of compact gearless generator designs became possible through rare earth magnets, specifically NdFeB and SmCo, because they provide sufficient magnetic flux density for offshore and large-scale wind installations. Wind turbine systems used to require geared driving systems that boosted blade rotational slowness with gearboxes to operate compatible with generator needs for high speeds (Sharma et al., 2019c). Gearboxes require complex mechanical construction while experiencing wear which leads to expensive operational failures and maintenance costs especially in offshore or remote applications. Direct-drive systems have proven to be important because they eliminate the requirement for gearboxes which leads to better turbine reliability while making system architecture simpler. High-performance rare earth permanent magnets have enabled generator rotor improvements that generate strong magnetic fields at low rotational speeds through their development and integration process (Sharma et al., 2022).

The innovation in power generation relies heavily on PMSGs. The generators incorporate rare earth magnets as rotor components that produce an unchanging magnetic field to energize stator windings into generating electricity (Srikant et al., 2021). The implementation of PMSGs produces advantages that surpass regular wound-rotor generators by delivering elevated efficiency (3–5% improvement) together with superior power factor and lowered copper losses and enhanced part-load characteristics.

Table 2 Key Rare Earth Elements used in Wind Energy and their Properties.

Element	*Symbol*	*Common Compound Used*	*Key Properties*	*Applications in Wind Energy*
Neodymium	Nd	$Nd_2Fe_{14}B$	High magnetic strength, corrosion-sensitive	Core component in NdFeB magnets
Dysprosium	Dy	Dy-doped NdFeB	Increases coercivity at high temps	Used in offshore turbine magnets
Praseodymium	Pr	(Nd,Pr)FeB alloys	Improves magnet performance	Substitute for Nd in magnets

Contd...

Table 2 *Contd...*

Element	*Symbol*	*Common Compound Used*	*Key Properties*	*Applications in Wind Energy*
Samarium	Sm	SmCo	Excellent thermal stability, corrosion-resistant	Used in SmCo magnets for extreme conditions
Terbium	Tb	Tb-doped NdFeB	Enhances high-temperature performance	Limited use due to high cost

Both reliability levels and electrical losses decrease because of complete removal of excitation systems and slip rings. Under offshore wind conditions the system benefits lead to both reduced Levelized Cost of Energy (LCOE) and increased operational lifespan because maintenance remains restricted and expensive. Table 2 represents the key REEs used in wind energy and their properties (Sharma et al., 2024a; Sharma et al., 2021a; Patel et al., 2024; Patel et al., 2025a). Rare earth magnet-based PMSGs enable the development of bigger wind turbine systems because they lead to increased energy capture potential while improving economies of scale. Ratings beyond 10 MW have led to the deployment of offshore wind farms where numerous direct-drive PMSG configurations appear (Patel et al., 2025b). NdFeB magnets enable wind power systems to create large torque effects at reduced speeds without requiring oversized or heavy generators. The offshore environment demands careful weight management and NdFeB magnets provide a compelling solution since they reduce the excessive mass that restricts tower safety and installation procedures. Floating offshore wind turbines have become important because they enable the extraction of wind power from deep water territories which cannot accommodate fixed-bottom installations. Because of platform dynamics and minimal service access and tough maritime elements these systems need small and durable generator solutions. Rare earth magnet-based generators demonstrate perfect suitability for demanded applications because they offer high power density characteristics alongside minimal maintenance needs. SmCo magnets remain underutilized because of their higher cost yet provide valuable features that include resistance against corrosion along with thermal stability which allow them to succeed in extreme floating wind conditions (Sharma et al., 2019d). The decentralized power market benefits from rare earth magnet utilization in small wind systems while commercial turbines continue to use these magnets for their operation. Table 3 represents the comparison of rare earth magnet recycling techniques (Sharma et al., 2020a). These systems prioritize reliability together with compactness and simple integration because they serve areas with insufficient infrastructure including rural and developing regions. Grants efficient power generation capabilities to permanent magnet generators built from NdFeB or SmCo materials so that wind energy reaches wider sections of the population.

Table 3 Comparison of Rare Earth Magnet Recycling Techniques.

Method	*Description*	*Advantages*	*Challenges*
Direct Reuse	Reusing magnets without altering structure	Low energy input, cost-effective	Limited by magnet condition and contamination
Hydrogen Decrepitation (HD)	Uses hydrogen to break down magnets into powder	Retains magnetic properties, scalable	Needs controlled conditions and purity
Solvent Extraction	Separates REEs using selective solvents	High purity, can handle mixed streams	Toxicity, chemical waste
Electrochemical Recovery	Uses electrolysis for separation of REEs	Environmentally friendlier, lower chemical load	Still in experimental stage
Bioleaching	Uses bacteria or fungi to leach REEs	Low environmental impact	Slow process, low efficiency currently

These applications benefit most from generator capabilities of producing power through minimum mechanics at lower cut-in speeds. Manufacturing and transport and assembly processes of modular generators became easier because of the implementation of rare earth magnetic materials. The benefit of modular design aligns perfectly with offshore high-capacity turbines and matches the ongoing movement to create industrial-scale and scalable wind farm developments. Each rare earth-based module provides high density power output together with structural stability and sufficient thermal capacity functionality through its magnetic attributes (Sharma et al., 2020a; Sharma et al., 2020b; Sharma et al., 2024b; Sharma et al., 2021b; Kaya et al., 2021). The modular PMSG configuration in numerous designs minimizes each component size leading to better logistics support and easier maintenance activities. Even though the benefits of integrating rare earth magnets in wind generator systems are evident the implementation comes with its own set of obstacles. The increase in wind turbine scale and number drives up the consumption of magnets made from neodymium, dysprosium and samarium among other rare earth materials. The scenario creates substantial supply chain challenges because of the restricted geographical distribution of critical material producing and refining activities (Kaya, 2024). In order to lessen these concerns, researchers together with manufacturers examine generator designs that optimize magnetic material use and incorporate replaceable material options that are common and recyclable. The study of hybrid generators with ferrite magnets along with variable flux designs continues but their performance does not meet the levels achieved by systems based on rare earth materials. Magnet protection together with thermal management stand as essential challenges for generator design developers. Conditions of extreme temperatures that affect wind turbines during operation including ambient temperatures and heat production result in magnet demagnetization leading to system efficiency downturns. The integration of effective cooling systems together with advanced thermal barrier materials along with protective coatings within generator assemblies enables magnet performance preservation to achieve long-term reliability (Khan et al., 2025). Generator designers now use numerical simulation tools and digital twin models to project thermal demands and enhance generator construction before production. Figure 4 shows the distribution of rare earth magnet types in wind turbines.

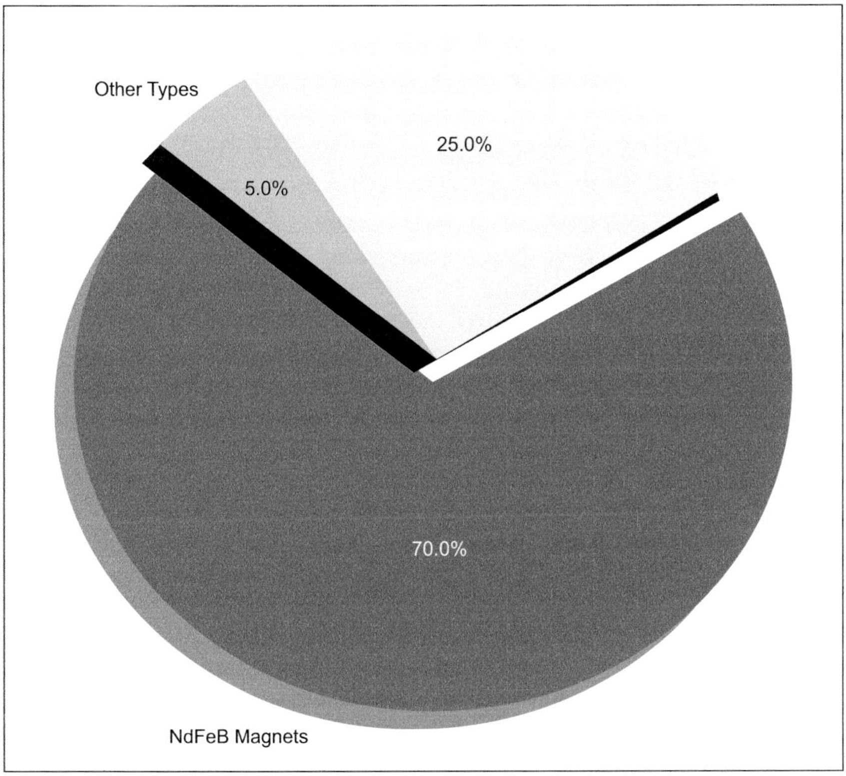

Fig. 4 Distribution of rare earth magnet types in wind turbines.

Future forecasts confirm rare earth magnets will maintain their essential role in wind turbine generator technological applications. Turbine manufacturers who pursue power generation increases while enhancing efficiency and reliability will drive up demand for advanced magnetic materials which will drive development of optimal generator configurations (Kaya et al., 2021).. The future of magnetic components and sustainability is supported by additive manufacturing (3D printing) and magnet recycling together with high-entropy magnetic alloys. For the next-generation wind power equipment it will be essential to merge these innovations with continuous system surveillance combined with Artificial Intelligence-based component prediction and integrated grid functionality (Kaya et al., 2021).

Wind turbine generators utilize rare earth magnets in PMSGs to drive significant improvements in the contemporary wind energy system design as well as performance (Patel et al., 2024). The development of efficient direct-drive systems with compact dimensions and high reliability becomes possible through their application which meets specific requirements of offshore and large-scale projects. Research along with ongoing innovation actively works to resolve the existing challenges related to supply chain risks and material deterioration. The future of sustainable wind turbine technology depends on rare earth magnets as generators because their usage continues to increase its dominance in global power generation.

4. Challenges and Sustainability Concerns Associated with Rare Earth Magnets in Wind Energy

The implementation of rare earth magnets in wind energy systems has delivered amazing generator designs yet this approach produces multiple complicated issues alongside sustainability problems. Table 4 represents the global strategies for rare earth magnet supply chain diversification. Deliveries of rare earth magnets throughout their complete lifecycle starting from raw material extraction to end-of-life management creates extensive environmental consequences which threaten the permanent sustainability of wind power generation technologies (Bailey et al., 2017; Balaram, 2019; Binnemans et al., 2013; Binnemans et al., 2015). The rapid global shift towards renewable energy makes it crucial to solve environmental and economic and geopolitical problems linked with rare earth magnets in wind turbine generators which face increasing demand from high-capacity wind projects.

The extraction of REEs which include neodymium, dysprosium, terbium and samarium along with their mining serves as the main obstacle in REE retrieval. Research classifies these elements as rare despite their crustal prevalence because they appear in small profit-making deposits which demand strong chemical extraction methods for purification (Desalegn et al., 2023) Large amounts of dangerous radioactive waste form during the processing of the common rare earth sources bastnäsite, monazite, and xenotime. Thorium represents a substantial environmental risk when it appears naturally near rare earth deposits because this radioactive element leads to significant hazards in improper waste management settings. Waste management mistakes at extraction locations generate extensive damage to groundwater supplies and destroy animal habitats and permanently endanger human health especially in nations whose environmental restrictions are either inadequately implemented or completely non-existent (Gutfleish et al., 2011).

The production of REEs takes place primarily in a single geographical area because China possesses more than 80% control over both extraction and refinement operations. The dominance of China over the rare earth supply chain creates substantial geopolitical threats and delivery instability because global rare earth demand increases due to clean energy technology development and electric vehicle

Table 4 Global Strategies for Rare Earth Magnet Supply Chain Diversification.

Country/Region	*Strategy/Initiative*	*Objective*	*Status/Impact*
United States	Department of Energy + CMI + Inflation Reduction Act	Domestic REE mining, R&D, recycling	Active, funding and pilot plants ongoing
European Union	European Raw Materials Alliance (ERMA), CRM Act	Reduce import dependency, boost recycling	Regulation drafted, pilot projects launched
Japan	Investment in REE mining in Vietnam and Malaysia	Supply security and cost stability	Active, with commercial-scale production
Australia	Support for Lynas Corporation and other miners	Develop rare earth refining and export	Strong production and export capacity
India	Public-private R&D and coastal monazite sand extraction	Enhance domestic resource utilization	Early-stage industrial investment
China	Government control, vertical integration	Maintain global dominance in REE supply chain	Leading producer with strategic export control

manufacture alongside consumer electronics adoption. Wind turbine production becomes financially unstable due to trade barriers, export quotas and market manipulation in countries that source magnets from abroad. These restrictions delay manufacturing schedules while raising overall project costs (Hakkar et al., 2021). After China increased its trade control over rare earths, several countries started essential strategic moves to strengthen their supply chains by developing local rare earth deposits and forming business partnerships in Australia and the U.S. and African countries and investing in enhanced refining and recycling technology to decrease Chinese export dependence.

Economic volatility presents itself as a critical issue related to rare earth materials. Research shows how prices of neodymium and dysprosium undergo steep changes because policymakers alter markets or because of material shortages or excessive market speculation (Binnemans et al., 2015). The unpredictable price behavior challenges wind energy developers when planning their projects since it affects the ability to accurately forecast costs. To achieve stable costs, wind turbine manufacturers must choose between material stockpiling or extended supply contracts which creates budgetary and logistical tension. Higher magnet performance through the use of heavy REEs such as dysprosium has worsened the rare earth imbalance because heavy elements are rarer to extract and more scarce than light elements.

A technical matter involving magnet aging along with gradual performance degradation constitutes a key challenge. NdFeB magnets will lose their magnetic properties after reaching their Curie point unless they incorporate heavy rare earth stabilizing additives (BNEF, 2023). The magnetic performance of stabilized magnets remains robust but gets diminished through both extended temperature increases along with repeated temperature fluctuations in offshore environments with significant temperature contrasts. The susceptibility to corrosion affects rare earth magnets continuously. The high moisture sensitivity of NdFeB magnets creates conditions which lead to performance-compromising pitting and cracking and flaking damage. Protective layers from nickel, epoxy, and parylene exist for risk reduction yet they increase equipment complexity and generate new potential failure areas within generator systems. The process of recycling used wind turbine magnets during their end-of-life phase remains poorly developed at a time when sustainability becomes increasingly important (Alves et al., 2020). The approaching end of service for initial massive wind turbines creates an immediate need to formulate proper recycling strategies for their rare earth magnets. The recycling of NdFeB magnets faces obstacles due to their tiny dimensions along with powerful magnetic properties and because adhesive or protective coatings hinder mechanical and chemical treatment processes. Current NdFeB magnet recycling operations utilize hydrogen decrepitation together with wet chemical extraction and pyrometallurgical techniques, yet all these methods are energy-expensive and have limited global implementation (Baghel & Nlebedim, 2023). Maintaining environmental sustainability in rare earth recycling requires the establishment of well-scalable methods combined with economic efficiency and environmental stewardship for achieving circular economic systems while addressing wind energy technology footprints. The social issues along with ethical aspects regarding rare earth mining operations face growing public evaluation. Rare earth mining operations that occur in developing world areas have triggered extensive displacement of land as well as significant labor exploitation and numerous community disputes. The social credibility and community support for wind energy projects depends critically on ethical raw material sourcing which must meet international guidelines including the OECD Due Diligence Guidance for Responsible Mineral Supply Chains. The industry has begun adopting certification programs and traceability systems along with third-party audits to verify responsible rare earth sourcing practices yet broader institutional adoption and enforcement weak points remain evident (Hakkar et al., 2021).

Significant research alongside development initiatives aim to decrease wind energy applications' dependence on critical rare earth materials to address multiple operational challenges. Researchers design magnets with low or no dysprosium content while utilizing material engineering techniques to maintain performance characteristics with reduced rare earth fractions and study alternative magnetic materials including alnico and ferrites and high-entropy alloys. Research continues to develop magnet options that cannot compete with rare earth magnetic energy density and performance but are being investigated for hybrid generator designs together with niche applications that can accept weight and efficiency trade-offs (Faisal et al., 2022).

5. Wind Energy Systems and Rare Earth Magnets: Extended Views

5.1 Comparative Performance of Rare Earth Magnet Generators

The process of integrating the rare earth magnets into the present wind turbine generators largely achieves its goal through the design and use of PMSGs. As for the variable-speed wind applications, these generators have shown better efficiency compared to the traditional doubly-fed induction generators (DFIGs) or the squirrel cage induction generators (SCIGs). Although DFIGs are still prevalent especially in onshore installations because of their maturity and cost-effectiveness, PMSGs provide performance pivotal advantages such as a wider speed operability range, higher torque density as well as elimination of gear boxes that are major failure and maintenance expenses (Binnemans et al., 2013).

Field comparisons show that systems using PMSGs can deliver full-load efficiencies scaling above 98%, DFIGs tend to be between 94–94%. Besides, PMSGs are quieter and they are more suitable for offshore wind farms where access for maintenance is limited and reliability becomes the key aspect (Binnemans et al., 2015) Installs in such harsh marine environments as the North Sea and coastal China have tested out rare earth-based generators for their codependence on long-term use, while corrosion and thermal issues require extravagant coatings, cooling, and air quality control.

5.2 Use of Rare Earth Magnets Economic and Life Cycle Analysis

A full economic evaluation of wind turbines with rare earth magnets should go beyond investment cost (i.e., capital expenditures, CAPEX), to include operation and maintenance costs (OPEX), levelized cost of energy (LCOE), or end-of-life value recovery. And while the initial cost of PMSGs is higher (because of the cost of NdFeB magnets), this cost is offset by lower OPEX over the wind turbine's 20–25-year lifespan (due to various and increased mechanical complexity and maintenance needs).

According to life cycle assessments (LCA), even though the environmental impact from rare earth extraction is high, the nature of reduced mechanical losses and increased power output from PMSGs over a system's lifecycle make it a viable choice for environmental sustainability (Bailey et al., 2017; Balaram, 2019; Binnemans et al., 2013; Binnemans et al., 2015; Binnemans et al., 2021; Blagoeva et al., 2016, BNEF, 2023). The net carbon footprint on a MWh generated is smaller when compared with other generator systems in terms of full lifecycle. Sustainability assessments made in the future are likely to consider the recycling potential and the utilization of rare earth materials as key indices.

5.3 Hybrid Generator Designs and the Minimised Rare Earth Dependency

In the bid to balance performance and resource consumption, hybrid generator configurations are attracting the eye. Such designs use both types of magnets, rare earth and ferrite or combine electromagnetic and permanent magnet excitement. For example, Ferrite-assisted synchronous reluctance machines (FASRMs) and hybrid excitation synchronous machines (HESMs) are becoming attractive alternatives (Chen et al., 2022).

The FASRMs lessen dependency on NdFeB while ensuring moderate torque and efficiency and are applicable to the mid-scale wind turbines. On the other hand, the HESMs provide controlled magnetic flux that can increase performance capabilities under a variable wind and provide fault tolerance. Similarly, the studies on these hybrid designs are also concentrating on scaling and manufacturability, in particular, with additive manufacturing techniques that enable complex magnetic circuit geometries with little wastage (Bollero & Polmero, 2022).

5.4 Geopolitical Strategies with National Security Implications

The rare earth supply chain is not only a technical question but also a question of national security and industrial competitiveness. A few governments have since realized the strategic danger of depending on a country that is a dominant supplier of essential materials, which at the moment is China. In the United States, the department of defense and the department of energy has put rare earths to be essential in national infrastructure, defense and clean energy. As such, rare earth exploration and separation facilities are being funded as well as magnet manufacturing all within the U.S. borders.

The EU, on its part, has also established the European Raw Materials Alliance (ERMA) and Critical Raw Materials Act to ensure a diversified supply chain and reduced environmental effects due to recycling and substitution. Japan, which experienced rare earth supply disruption in 2010, has made massive investments in alternative sources in South East Asia and technologies of magnet miniaturization. The national strategies are supported by international arrangements such as the U.S.-Australia critical minerals partnership and the Quad Rare Earth Coordination Network (U.S., Japan, India, and Australia).

5.5 Academic Research and Industrial Collaborations

More and more cooperative studies are aimed at the sustainability of rare earth magnet, sponsored by the state and private companies. Examples include:

REECOVER Project (EU): Directed towards recycling the permanent magnets from end-of-life products by environmentally friendly procedures (Ali et al., 2017).

CAMP (Critical Materials Institute, U.S.): The investigation of the substitutes for the rare earths, enhancement of the efficiency of the magnetic materials.

Magnet4Europe: A consortium that envisions to minimize the use of dysprosium and terbium in high-performance magnets and also develop new sintering and diffusion technologies (Bailey et al., 2017).

Such collaborations promote knowledge sharing among universities, industrial as well as policymaking partners, making sure that innovation at lab-scale is transferred to commercial-scale applications (Archipenko et al., 2023).

5.6 Rare Earth-Free Alternatives and Emerging Technologies

Nonetheless, great efforts are underway to see how other magnets, that are non-rare earths, can be developed to prevent rare earths from dominating the wind generator industry. Two prominent candidates are:

Iron-Nitride ($Fe_{16}N_2$) Magnets: $Fe_{16}N_2$ has a high level of magnetization and also the possibility of rare earth-free high-performance magnets; its metastability and large-scale synthesis difficulties keep it as the research interest.

MnBi and Co-based Alloys: These compounds promise more economical and better corrosion-resistant magnetic properties, which, however, are not so efficient compared to the efficiency of NdFeB.

Also, superconducting generators, although too costly and complex at present, provide pure theoretical prospect of eliminating conventional magnetic materials in addition to providing outrageous power densities.

5.7 Environmental Justice and Community Engagement

Vicinities of rare earth mining operations, especially in the Global South, tend to experience unfair environmental impacts. The pollution heritage in China's

Bayan Obo, together with the illegal mining in Myanmar, emphasizes the necessity for establishing the environmental justice frameworks. Any future wind energy projects should guarantee open and honorable sourcing of rare earth materials, inclusive of all stakeholders, benefit sharing, and FPIC practices (Coey, 2020; Croat & Ormerod, 2022; Cui et al., 2018; Desalegn et al., 2023; Dong et al., 2021).

SIAs are being more and more included in supply chain certification and wind energy businesses are to contribute to developmental programs in origin areas. Embracing traceability systems in blockchain and international standards for reporting such as GRI (Global Reporting Initiative) will increase accountability across the rare earth supply chain.

5.8 Educational and Workforce Development Implications

The same investment in human capital is needed in order to achieve the shift to energy technologies intensive in rare earths. Universities and technical institutes need to review curricula to include multidisciplinary subjects including rare earth geochemistry; magnetics; green extraction technologies; and principles of circular economy. In order to create an expertise in the manufacturing of magnets, and recycling, development programs in the workforce are needed (Binnemas et al., 2015; Binnemans et al., 2021; Blagoeva et al., 2016; BNEF, 2023; Bollero & Polmero, 2022). Development for system level integration is also necessary. Scholarships, internships, and vocational training in this field are starting to be offered by governments and industry associations which acknowledge its relevancy to the future infrastructure of clean energy.

6. Future Trends and Conclusion

During the current global energy sector transition toward sustainability and decarbonization wind energy proves essential for forming a diverse renewable energy portfolio. Wind turbine generators receive performance and efficiency improvements and enhanced reliability through integration of NdFeB and SmCo rare earth magnets in their systems which operate within onshore and offshore wind projects. Previous sections have shown how increased dependence on essential materials generates sustainability issues with overlapping economic and geopolitical elements. The long-term success of wind energy applications using rare earth magnets depends on correct predictions about material innovation combined with design modifications and waste reduction techniques alongside government framework development.

Next-generation magnetic materials become a significant leading trend in future technological advancements. Scientists continue to research novel material systems which either lessen REE content or remove it entirely while maintaining magnetic properties. Grain boundary diffusion technology shows promise to minimize heavy rare earth elements such as dysprosium in magnets through precise magnetic performance enhancement. High-entropy alloys and nanocomposite magnets show potential as future magnetic materials through their tunable properties combined with enhanced stability at high temperatures. The new material innovations focus on three key areas: supply pressure reduction alongside cost reduction and material adaptation for difficult operational settings like floating wind turbines.

Design and engineering developments will modify the function of rare earth magnets within wind turbine systems. Next-generation wind turbines will adopt transverse flux or axial flux generator topologies because these designs enable increased torque density along with reduced mass while generating the same level of power. The exact design of magnet placement and magnetic circuit geometry together with thermal pathways becomes possible with advanced modeling methods that integrate machine learning optimization and digital twin simulations which boost material efficiency and operational lifespan. Combining innovative developments with intelligent condition monitoring technology enables operators to obtain immediate performance assessments and schedule proactive maintenance initiatives which in turn shortens operational disruptions and enhances investment returns.

The recycling sector together with the circular economy framework for rare earth magnets continues to form an essential part of future market development. Thousands of wind turbine installations will reach their end-of-life stage during future decades which will create massive opportunities to recover the contained in their magnet systems. Plant-based recycling methods under development combine magnet powder reuse techniques with electrochemical methods alongside green hydrometallurgical techniques to recover rare earth elements while minimizing both energy consumption and environmental impact during the extraction process. Efficient circular systems help decrease the need for fresh materials and decrease mine-caused environmental destruction while establishing seamless supply networks. Independent policy tools such as Extended Producer Responsibility and resource recovery incentives will become prominent factors in advancing proper end-of-life practices for components containing REEs. The rare earth supply chains have experienced both geopolitical expansion and stronger infrastructure developments. The global push toward rare earth production and processing independence led the United States and Australia along with Canada and members of the European Union to build domestic mining capabilities and refining capacities and magnet manufacturing abilities. Through strategic alliances and rare earth stockpiling along with public-private partnerships organizations seek to create buffers against supply disruptions and price volatility. Establishing transparent supply chains with ethical sourcing practices and traceability features represents a critical path for wind energy to sustain its credibility while delivering resilient and scalable growth as demand exceeds expectations. Wind energy systems will benefit from changing policy and regulatory frameworks that develop pathways to successfully incorporate rare earth magnets. The European Green Deal and The U.S. Inflation Reduction Act together with the Critical Raw Materials Act are currently driving investments in renewable technology and material supply chain resilience through international initiatives. These policies both advance research initiatives and require all sectors to implement superior sourcing methods alongside production techniques and recycling procedures. The deployment of responsible magnet utilization and supply requires both consistent border regulations together with clear reporting standards for global industry cooperation. Rare earth magnets, integrated into new wind energy concepts, will drive future innovation advancements. The future generation of renewable energy systems including floating wind farms and airborne wind energy systems and vertical-axis turbines and hybrid renewable systems (wind-solar-storage integration) will necessitate innovative generator technologies that combine lightweight design with compact size and elimination of maintenance needs. Rare earth magnets demonstrate excellent stability and high power density properties for new applications but necessary challenges still need resolution. Wind energy systems will evolve through a combination of advanced materials, smart electronics, artificial intelligence, and sustainability frameworks to become both technologically advanced and environmentally friendly installations.

The advancements in modern wind energy depend heavily on rare earth magnets for developing efficient direct-drive systems with increased capacity performance. The exceptional magnetic characteristics of these magnets enable generator systems to reach their most compact design while delivering best-in-class performance combined with reduced mechanical requirements. The escalating usage of REEs creates fundamental issues regarding sustainable environmental practices, limited resource stocks, and secure supply networks together with ethical mining procedure standards. A solution requires global cooperation involving innovation from material sciences with circular economy approaches and new policies together with international agreements and rules. Wind energy's future with rare earth magnets relies on both technological progress and successful stewardship of these vital materials through careful environmental and social practices. The wind energy sector's ability to keep advancing as foundational to global clean energy transitions depends on proactive solutions to minimize these concerns which will yield lasting environmental and economic advantages.

References

Adamas Intelligence (2020). 2624 tonnes of rare earth oxides and metals were consumed globally in the production of traction motors for EVs in 2019. Accessed 28 November 2023.

Adamas Intelligence (2023). The skyrocketing value of rare earths powering the energy transition. Accessed 28 November 2023.

Ali, S.H., Giurco, D., Arndt, N., Nickless, E., Brown, G., Demetriades, A., Durrheim, R., Enriquez, M.A., Kinnaird, J., Littleboy, A. and Meinert, L.D. (2017). Mineral supply for sustainable development requires resource governance. *Nature*, 543(7645), 367–372. https://doi.org/10.1038/nature21359.

Alves Dias, P., Bobba, S., Carrara, S. and Plazzotta, B. (2020). *The Role of Rare Earth Elements in Wind Energy and Electric Mobility*. EUR 30488 EN, Publication Office of the European Union, Luxembourg. https://doi.org/10.2760/303258.

Arkhipenko, A.A., Marina, G.E., Doronina, M.S., Korotkova, N.A. and Baranovskaya, V.B. (2023). X-ray fluorescence analysis of waste Sm-Co magnets: A rational approach. *Recycling*, 8, 84. https://doi.org/10.3390/recycling8060084.

Arnold Magnetic Technologies. (2023). 3 key parameters to consider when choosing a permanent magnet. Arnold Magnetic Technologies. Accessed 19 November 2023.

Baghel, A.P.S., & Nlebedim, I.C.A. (2023). Hybrid rotor design with reduced rare earth magnets for traction motors. *JOM*, *75*, 557–565. https://doi.org/10.1007/s11837-022-05594-5.

Bailey, G., Mancheri, N. and Van Acker, K. (2017). Sustainability of permanent rare earth magnet motors in (H)EV industry. *J. Sustain. Metall.*, 3, 611–626. https://doi.org/ 10.1007/s40831-017-0118-4.

Balaram, V. (2019). Rare earth elements: A review of applications, occurrence, exploration, analysis, recycling, and environmental impact. *Geosci. Front.*, 10(4), 1285–1303. https://doi.org/10.1016/j.gsf.2018.12.005.

Binnemans, K., Jones, P.T., Blanpain, B., Van Gerven, T., Yang, Y., Walton, A. and Buchert, M. (2013). Recycling of rare earths: A critical review. *J. Clean. Prod.*, 51, 1–22. https://doi.org/10.1016/j.jclepro.2012.12.037.

Binnemans, K., Jones, P.T., Blanpain, B., Van Gerven, T. and Pontikes, Y. (2015).Towards zero-waste valorisation of rare-earth-containing industrial process residues: A critical review. *J. Clean. Prod.*, 99, 17–38. https://doi.org/10.1016/j.jclepro.2015.02.089.

Binnemans, K., McGuiness, P. and Jones, P.T. (2021). Rare-earth recycling needs market intervention. *Nat. Rev. Mater.*, 6(6), 459–461. https://doi.org/10.1038/s41578-021-00308-w.

Blagoeva, D., Alves Dias, P., Marmier, A. and Pavel, C.C. (2016). *Assessment of the Potential Bottlenecks Along the Materials Supply Chain for the Future Deployment of Low Carbon Energy and Transport Technologies in the EU*. Joint Research Centre (JRC). Publications Office of the European Union. EUR 28192 EN.

BNEF. (2023). *Electric Vehicle Outlook 2023*. Bloomberg New Energy Finance. Accessed 14 April 2024.

Bollero, A. and Palmero, E.M. (2022). Recent advances in hard-ferrite magnets. In: Croat, J., Ormerod, J. (Eds.), *Modern Permanent Magnets*, pp. 65–112. Woodhead Publishing Series in Electronic and Optical Materials. https://doi.org/10.1016/B978-0-323-88658-1.00013-3.

Chang, C.C.W., Ding, T.J., Ping, T.J., Chao, K.C. and Bhuiyan, M.A.S. (2022). Getting more from the wind: Recent advancements and challenges in generators' development for wind turbines. *Sustain. Energy Technol. Assess.*, 53, 102731. https://doi.org/10.1016/j.seta.2022.102731.

Chen, Z., Li, Z., Chen, J., Kallem, P., Banat, F. and Qiu, H. (2022). Recent advances in selective separation technologies of rare earth elements: A review. *J. Environ. Chem. Eng.*, 10(1), 107104. https://doi.org/10.1016/j.jece.2021.107104.

Coey, J.M.D. (2020). Perspective and prospects for rare earth permanent magnets. *Eng. J.*, 6(2), 119–131. https://doi.org/10.1016/j.eng.2018.11.034.

Croat, J.J. and Ormerod, J. (2022). The history of permanent magnets. In: Croat, J., Ormerod, J. (Eds.), *Modern Permanent Magnets*, pp. 1–30. Woodhead Publishing Series in Electronic and Optical Materials. https://doi.org/10.1016/B978-0-323-88658-1.00004-2.

Cui, J., Kramer, M., Zhou, L., Liu, F., Gabay, A., Hadjipanayis, G., Balasubramanian, B. and Sellmyer, D. (2018). Current progress and future challenges in rare-earth-free permanent magnets. *Acta Mater.*, 158, 118–137. https://doi.org/10.1016/j.actamat.2018.07.049.

Desalegn, B., Gebeyehu, D., Tamrat, B., Tadiwose, T. and Lata, A. (2023). Onshore versus offshore wind power trends and recent study practices in modeling of wind turbines' life-cycle impact assessments. *Clean. Eng. Tech.*, 17, 100691. https://doi.org/10.1016/j.clet.2023.100691.

Dong, Z., Mattocks, J.A., Deblonde, G.J.P., Hu, D., Jiao, Y., Cotruvo Jr, J.A. and Park, D.M. (2021). Bridging hydrometallurgy and biochemistry: A protein-based process for recovery and separation of rare earth elements. *ACS Cent. Sci.*, 7(11), 1798–1808. https://doi.org/10.1021/acscentsci.1c00724.

Dow, J. (2023). *Tesla is Going (back) to EV Motors with no Rare Earth Elements*. Electrek.

Faisal, L., Rama, V.S.B., Yang, J.M., Wajid, A. and Ghorui, S.K. (2022). Performance and simulation analysis of IPMSyncRM (internal permanent magnet synchronous reluctance motor) for advanced electric vehicle design. In: *2022 3rd International Conference for Emerging Technology* (*INCET*). *IEEE*, pp. 1–6. https://doi.org/10.1109/INCET54531.2022.9824716.

Ford. (2021). Ford Europe goes all-in on EVs on road to sustainable profitability; cologne site begins $1 billion transformation. Accessed 28 November 2023.

Fujita, Y., McCall, S.K. and Ginosar, D. (2022). Recycling rare earths: Perspectives and recent advances. *MRS Bull.*, 47, 283–288. https://doi.org/10.1557/s43577-022-00301-w.

Gao, W., Kang, X., Zhang, J. and Chi, Z. (2022). Determining the maximum operating temperature of permanent magnets via *in-situ* measurement of the magnetic flux. *J. Magn. Magn. Mater.*, 562, 169744. https://doi.org/10.1016/j.jmmm.2022.169744.

Ghorbani, Y., Zhang, S.E., Nwaila, G.T., Bourdeau, J.E. and Rose, D.H. (2023). Embracing a diverse approach to a globally inclusive green energy transition: Moving beyond decarbonisation and recognising realistic carbon reduction strategies. *J. Clean. Prod.*, 140414. https://doi.org/10.1016/j.jclepro.2023.140414.

Ghorbani, Y., Zhang, S.E., Bourdeau, J.E., Chipangamate, N.S., Rose, D.H., Valodia, I. and Nwaila, G.T. (2024). The strategic role of lithium in the green energy transition: Towards an OPEC-style framework for green energy-mineral exporting countries (GEMEC). *Resour. Policy*, 90, 104737. https://doi.org/10.1016/j.resourpol.2024.104737.

Ghorbani, Y., et al. (2025). Rare earth permanent magnets for the green energy transition: Bottlenecks, current developments, and cleaner production solutions. *Resources, Conservation, and Recycling*, *212*(7645), 10796613.

Gielen, D. and Lyons, M. (2022). *Critical Materials for the Energy Transition: Rare Earth Elements*. International Renewable Energy Agency (IRENA), Abu Dhabi. Accessed 14 April 2024.

Global Wind Energy Council (GWEC). (2023). *Global Wind Report 2023*. Accessed 28 November 2023.

Government of Canada. (2022). *The Canadian Critical Minerals Strategy: From Exploration to Recycling*.

Government of Canada, Natural Resources Canada.Gutfleisch, O., Willard, M.A., Brück, E., Chen, C.H., Sankar, S.G. and Liu, J.P. (2011).Magnetic materials and devices for the 21st century: Stronger, lighter, and more energy efficient. *Adv. Mater.*, 23(7), 821–842. https://doi.org/10.1002/adma.201002180.

Hakkar, M., Arhouni, F.E., Mahrou, A., Bilal, E., Bertau, M., Roy, A., Steiner, G., Haneklaus, N., Mazouz, H., Boukhair, A. and Benjelloun, M. (2021). Enhancing rare earth element transfer from phosphate rock to phosphoric acid using an inexpensive fly ash additive. *Miner. Eng.*, 172, 107166. https://doi.org/10.1016/j.mineng.2021.107166.

Harris, I.R. and Jewell, G.W. (2012). Rare-earth magnets: Properties, processing and applications. In: Kilner, J.A., Skinner, S.J., Irvine, S.J.C., Edwards, P.P. (Eds.), *Functional Materials for Sustainable Energy Applications*, pp. 600–639. Woodhead Publishing Series in Energy. https://doi.org/10.1533/9780857096371.4.600.

Kaya, E.E., Kaya, O., Stopic, S., Gürmen, S. and Friedrich, B. (2021). NdFeB magnets recycling process: An alternative method to produce mixed rare earth oxide from scrap NdFeB magnets. *Metals* (*Basel*). 11, 716. https://doi.org/10.3390/met11050716.

Kaya, M. (2024). An overview of NdFeB magnets recycling technologies. *Current Opinion Green Sustain. Chem.*, 46, 100884. https://doi.org/10.1016/j.cogsc.2024.100884.

Khan, F.N.U., Rasul, M.G., Sayem, A.S.M. and Mandal, N., (2023). Maximizing energy density of lithium-ion batteries for electric vehicles: A critical review. *Energy Rep.*, 9, 11–21. https://doi.org/10.1016/j.egyr.2023.08.069.

Patel, D.K., Sharma, V.K. and Sharma. (2024). Advancements in aluminum matrix composites: A comprehensive review on the mechanical and microstructural attributes of rare earth elements (REEs) and rare earth oxides (REOs) doped materials. *Transactions of the Indian Institute of Metals*, *77*(9), 2265–2276.

Patel, D.K., Sharma, V.K. and Sharma. (2025a). Current global Status of rare earth elements (REEs) and their role as catalysts in reducing air pollution for maintaining environmental sustainability. *Transactions of the Indian Institute of Metals*, *78*1), 1–11.

Patel, D., Sharma, V.K., Kumar, V., Kumar, P. and Kumar, D. (2025b). Enhancing hydrophobicity and anti-corrosion properties of Al-6061 Aluminum hybrid composites through trace additions of rare earth oxides. In: *Proceedings of the Institution of Mechanical Engineers, Part E: Journal of Process Mechanical Engineering*, 09544089251317381.

Sharma, V.K., Kumar, V. and Joshi, R.S. (2019a). Investigation of rare earth particulate on tribological and mechanical properties of Al-6061 alloy composites for aerospace application. *Journal of Materials Research and Technology*, 8(4), 3504–3516.

Sharma, V. K., Kumar, V. and Joshi, R.S. (2019b). Effect of RE addition on wear behavior of an Al-6061 based hybrid composite. *Wear*, 426, 961–974.

Sharma, V.K., Kumar, V. and Joshi, R.S. (2019c). Experimental investigation on effect of RE oxides addition on tribological and mechanical properties of Al-6063 based hybrid composites. *Materials Research Express*, 6(8), 0865d7.

Sharma, V.K., Kumar, V. and Kumar, V. (2019d).: Development of rare-earth oxides based hybrid AMCs reinforced with SiC/Al2O3: Mechanical and metallurgical characterization. *Journal of Materials Research and Technology*, 8(2), 1971–1981.

Sharma, V.K., Kumar, V. and Joshi, R.S. (2020a). Parametric study of aluminium-rare earth based composites with improved hydrophobicity using response surface method. *Journal of Materials Research and Technology*, 9(3), 4919–4932.

Sharma, V.K., Kumar, V., Joshi, R.S. and Sharma D. (2020b). Experimental analysis and characterization of SiC and RE oxides reinforced Al-6063 alloy based hybrid composites. *The International Journal of Advanced Manufacturing Technology*, 108, 1173–1187.

Sharma, V.K., Kumar, V. and Joshi, R.S. (2021a). Manufacturing of stable hydrophobic surface on rare-earth oxides aluminium hybrid composite. In: *Proceedings of the Institution of Mechanical Engineers, Part E: Journal of Process Mechanical Engineering*, 235(4), 899–912.

Sharma, V.K., Aggarwal D., Kumar V. and Joshi R.S. (2021b). Influence of rare earth particulate on the mechanical and tribological properties of Al-6063/SiC hybrid composites. *Particulate Science and Technology*, 39(8), 928–943.

Sharma, V.K., Kumar, V., Joshi, R.S., and Kumar A. (2022). Effect of REOs on tribological behavior of aluminum hybrid composites using ANN. In: *Additive Manufacturing in Industry 4.0*, pp. 153–167. CRC Press.

Sharma, V.K., Kumar, P., Akhai, S., Kumar, V. and Joshi, R.S. (2024a). Analyzing the tribological and mechanical performance of Al-6061 with rare earth oxides: An experimental analysis. In: *Proceedings of the Institution of Mechanical Engineers, Part E: Journal of Process Mechanical Engineering*, *238* (5), 2278–2292.

Sharma, V.K., Kumar P., Kumar D., Chalisgaonkar R. and Akhai S. (2024b). Characterization of Al 6061/Al2O3/SiC composites with cerium oxide: Corrosion analysis and microstructural insights. *Composites Theory and Practice*, 24.

Srikant, Sharma V.K., Kumar, V., Joshi R.S., Jain, S. and Kumar P. (2021). A review of recent research on rare earth particulate composite materials and structures with their applications. *Transactions of the Indian Institute of Metals*, 1–13.

14 Enhancing Wear Resistance and Structural Performance of Y_2O_3-Reinforced Al-6063 Composites for Energy-Efficient Applications

Dinesh Kumar[1*], Pardeep Kumar[2], Mahiti Gupta[3], Vipin Kumar Sharma[4], Mohamed Arezki Mellal[5] and Santosh Kumar Rai[6]

1. Introduction

The increasing demand for high-performance materials in the automotive industry has driven significant research into developing advanced composites with superior mechanical and tribological properties. Among the various materials available, aluminum alloys are widely used due to their lightweight, excellent thermal conductivity, and good corrosion resistance (Kumar, 2023) Specifically, Al-6063, a medium-strength aluminum alloy, has garnered attention for its applications in automotive components, including piston cylinder liners (Kumar et al., 2024). However, the inherent wear resistance of Al-6063 is often inadequate for high-stress applications, necessitating the development of reinforced composites to enhance its wear performance (Kumar et al., 2023). The piston cylinder liner arrangement in internal combustion engines is subjected to severe operational conditions, including high temperatures, pressures, and mechanical stresses (Kumar et al., 2022). These conditions lead to significant wear and tear, affecting the efficiency and lifespan of the engine. Therefore, improving the wear resistance of materials used in piston cylinder liners is crucial for enhancing engine performance and durability (Kumar et al., 2023). Reinforcing aluminum alloys with hard ceramic particles has emerged as a promising approach to address this challenge. Yttria oxide (Y_2O_3), a rare earth oxide, has gained prominence as a reinforcement material due to its exceptional

[1,2] Department of Mechanical Engineering, MMEC, Maharishi Markandeswar (Deemed to be University), Mullana-Ambala, India.
[3] Department of Bio-Sciences & Technology, MMEC, Maharishi Markandeswar (Deemed to be University), Mullana-Ambala, India.
[4] Department of Mechanical Engineering, Meerut Institute of Engineering & Technology, Meerut, U.P., India.
[5] LMSS, Department of Mechanical Engineering, Faculty of Technology, M'Hamed Bougara University, Boumerdes, Algeria.
[6] Department of Mechanical Engineering, Faculty of Engineering and Technology, SRM Institute of Science and Technology, Delhi NCR Campus, Ghaziabad, U.P., India.
* Corresponding author: dinesh_kumar@mmumullana.org

DOI: 10.1201/9781003683254-14

hardness, thermal stability, and chemical inertness (Kumar et al., 2023). The incorporation of yttria oxide particles into aluminum alloys can potentially enhance their wear resistance by providing a hard phase that impedes material loss during sliding contact (Singh & Angra, 2022). Despite its potential, limited research has been conducted on the wear behavior of Al-6063 alloy reinforced with yttria oxide, particularly in the context of piston cylinder liner applications (Kumar et al., 2023). This study aims to investigate the wear behavior of Al-6063 alloy reinforced with varying weight percentages of yttria oxide particles (0, 2, 4, and 6 wt.%) and to determine the optimal composition for use in piston cylinder liners. The specific objectives of the research are:

- To synthesize Al-6063/yttria oxide composites using the stir casting method, ensuring a uniform distribution of reinforcement particles within the matrix.
- To conduct a comprehensive microstructural analysis of the composites to evaluate the dispersion of yttria oxide particles and their impact on the matrix.
- To perform wear tests using a pin-on-disc apparatus under different loads and sliding speeds to assess the wear resistance of the composites.

Previous studies have demonstrated the benefits of reinforcing aluminum alloys with ceramic particles to enhance their mechanical and tribological properties. For instance, SiC (Kumar et al., 2023), Al_2O_3 (Kumar & Kumar, 2024), and TiC (Kumar et al., 2024) particles have been successfully incorporated into aluminum matrices, resulting in improved hardness, tensile strength, and wear resistance. These reinforcements create a hard phase within the composite, which acts as a barrier to dislocation movement and material loss during wear. Yttria oxide, in particular, has shown promise as a reinforcement material due to its unique properties. Studies on Al-Y_2O_3 composites have reported significant improvements in hardness and wear resistance, attributed to the hard and thermally stable nature of yttria oxide particles. However, the application of Y_2O_3 as a reinforcement in Al-6063 for specific automotive components like piston cylinder liners remains relatively unexplored.

2. Materials and Method

In this study, the primary materials used are Al-6063 alloy and yttria oxide (Y_2O_3) particles. The synthesis of Al-6063/Y_2O_3 composites involves the combination of these materials to enhance the wear resistance and mechanical properties of the resulting composite. Al-6063 is a medium-strength aluminum alloy with good extrudability and mechanical properties, making it suitable for a variety of structural applications. The chemical composition of Al-6063 is presented in Table 1. Al-6063 is selected for its balance of strength, ductility, and corrosion resistance, which are essential properties for applications subjected to dynamic and harsh operating conditions, such as piston cylinder liners.

Table 1 Chemical Composition of Al-6063 Alloy.

Element	*Composition (wt.%)*	*Element*	*Composition (wt.%)*
Aluminum (Al)	97.85	Manganese (Mn)	0.14
Magnesium (Mg)	0.7	Chromium (Cr)	0.16
Silicon (Si)	0.5	Zinc (Zn)	0.11
Iron (Fe)	0.15	Titanium (Ti)	0.10
Copper (Cu)	0.12	Other Elements	0.17

Yttria oxide (Y_2O_3) is a rare earth oxide known for its high hardness, thermal stability, and chemical inertness. These properties make Y_2O_3 an ideal reinforcement material for improving the wear resistance and mechanical performance of metal matrix composites. The specific properties of yttria oxide are summarized in Table 2.

Table 2 Properties of Yttria Oxide (Y2O3).

Properties	*Value*
Density	5.01 g/cm^3
Melting Point	2,450°C
Hardness	8 GPa
Thermal Conductivity	12 W/m·K

2.1 Stir-casting Approach

The Al-6063/yttria oxide composites will be synthesized using the stir casting method, which involves melting the Al-6063 alloy and adding yttria oxide particles to the molten metal (Kumar et al., 2023). The mixture will be stirred to ensure uniform dispersion of the particles, followed by casting into mould and subsequent solidification (Pardeep et al., 2023). The graphical abstract is shown in Figure 1.

The composites will be prepared with different weight percentages (Table 3) of yttria oxide (0, 2, 4, and 6 wt.%) to evaluate the effect of reinforcement content on wear behavior.

Table 3 Specimen designation and their composition.

S.No	*Designation*	*Composition*
1	CS1	100% Al-6063
2	CS2	98% Al-6063 + 2% Y_2O_3
3	CS3	96% Al-6063 + 4% Y_2O_3
4	CS4	94% Al-6063 + 6% Y_2O_3

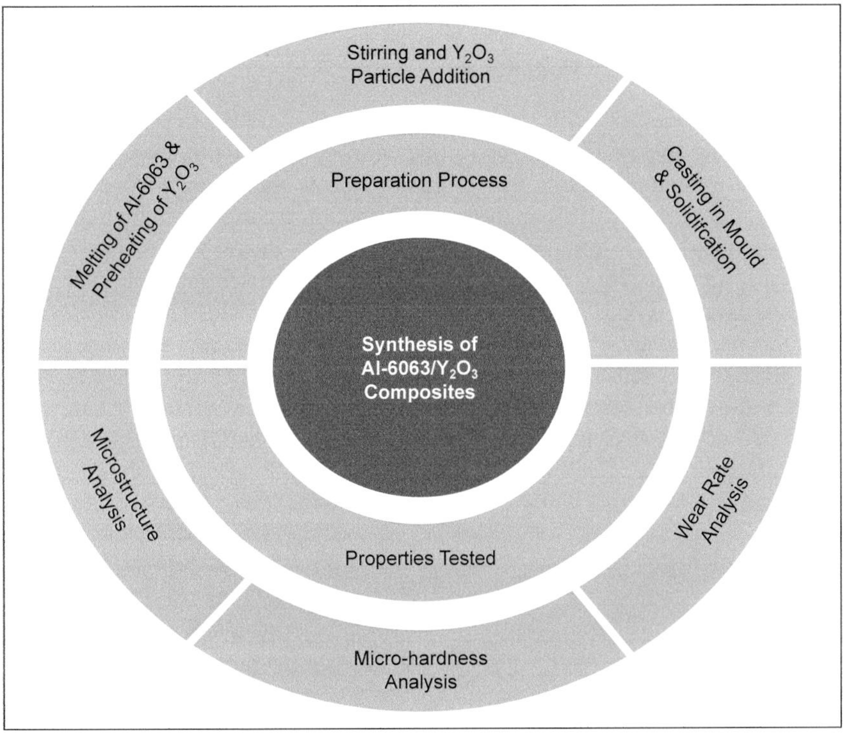

Fig. 1 Graphical representation of the synthesis of Al-6063/Y_2O_3.

2.2 Material Characterization

The microstructural analysis is performed using scanning electron microscopy (SEM) to examine the dispersion of Y_2O_3 particles within the Al-6063 matrix (Kumar & Singh, 2024). The digital Vickers hardness tester was used to measure the hardness of different fabricated specimens (Kumar et al., 2025). Wear tests are conducted using a pin-on-disc apparatus, where the composite pins are slid against a hardened steel disc under controlled conditions (Kumar et al., 2025). The parameters for wear testing include varying loads (10 N, 20 N, and 30 N) and sliding speeds (0.5 m/s, 1.0 m/s, and 1.5 m/s) to evaluate the wear resistance of the composites (Kumar et al., 2024). The wear rate is calculated by measuring the weight loss of the specimens after a fixed sliding distance (Kumar et al., 2024). The cast composite specimens are machined into standard dimensions for wear testing. The specimens are prepared in the form of pins with a diameter of 10 mm and a length of 30 mm (Patel & Sharma, 2025). The surface of each specimen is polished to achieve a smooth finish, ensuring consistent contact during wear testing (Kumar et al., 2025; Sharma et al., 2024; Sharma et al., 2023; Sharma et al., 2021; Shah et al., 2019; Singh et al., 2023).

3. Results and Discussions

3.1 Microstructural Results

The microstructural analysis of Al-6063/Y_2O_3 composites is crucial in understanding the distribution and interaction of yttria oxide (Y_2O_3) particles within the aluminum matrix. This section presents the findings from scanning electron microscopy (SEM) analyses conducted on the synthesized composites. SEM provides high-resolution images that allow for a detailed examination of the microstructure, including the morphology of Y_2O_3 particles and their interaction with the matrix. The SEM images show that Y_2O_3 particles have an irregular, angular morphology, contributing to their role in enhancing wear resistance. The particles are well-embedded in the Al-6063 matrix, indicating good bonding between the reinforcement and the matrix. The interface between the Y_2O_3 particles and the Al-6063 matrix appears to be clean and well-defined. There is no evidence of significant interfacial reactions or debonding, which could weaken the composite structure.

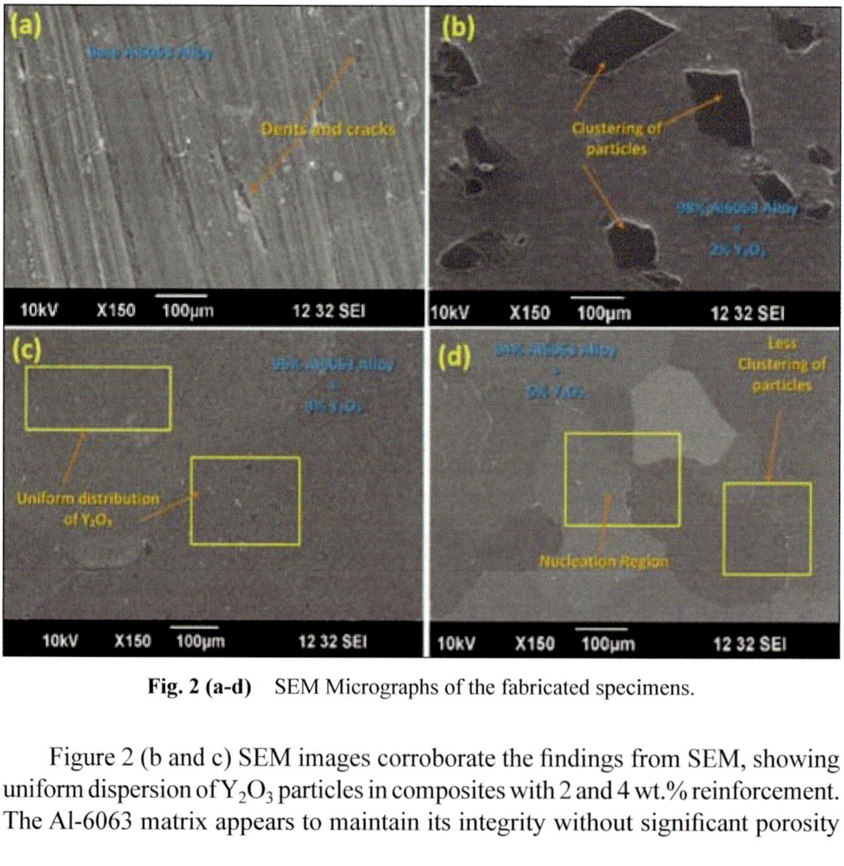

Fig. 2 (a-d) SEM Micrographs of the fabricated specimens.

Figure 2 (b and c) SEM images corroborate the findings from SEM, showing uniform dispersion of Y_2O_3 particles in composites with 2 and 4 wt.% reinforcement. The Al-6063 matrix appears to maintain its integrity without significant porosity or defects, indicating successful incorporation of Y_2O_3 particles during the stir-casting process. The OM images reveal that Y_2O_3 particles are relatively uniformly distributed throughout the Al-6063 matrix in all composite samples. This uniformity is essential for ensuring consistent mechanical and tribological properties. Some agglomeration is noted in the 6 wt.% composite, but it is not extensive. Figure (2d) of sample CS4 with higher Y_2O_3 content (6 wt.%), slight clustering of particles is observed. However, the degree of clustering is minimal and does not significantly affect the overall dispersion.

3.2 Hardness Analysis

Table 4 shows the Vickers hardness number (VHN) for different compositions of Al-6063 reinforced with yttria oxide (Y_2O_3).

Table 4 Hardness values of fabricated specimens.

S. No	*Designation*	*Composition*	*Hardness (VHN)*
1	CS1	100% Al-6063	68
2	CS2	98% Al-6063 + 2% Y_2O_3	81
3	CS3	96% Al-6063 + 4% Y_2O_3	104
4	CS4	94% Al-6063 + 6% Y_2O_3	97

The hardness of pure Al-6063 is 68 VHN, which serves as a baseline for comparing the reinforced composites. This relatively lower hardness is typical for unreinforced aluminum alloys, which are softer and more ductile.The addition of 2% Y_2O_3 increases the hardness to 81 VHN. This improvement is due to the dispersion of hard Y_2O_3 particles within the Al-6063 matrix. These particles hinder dislocation movement, making the composite harder and more resistant to deformation. With 4% Y_2O_3, the hardness jumps to 104 VHN, showing a substantial increase. This result indicates that the optimal dispersion of Y_2O_3 particles at this concentration significantly strengthens the composite. The particles effectively block more dislocations and enhance the load-bearing capacity of the matrix. Surprisingly, the hardness decreases to 97 VHN with 6% Y_2O_3, even though it is still higher than the unreinforced sample.

At higher concentrations, Y_2O_3 particles may start to cluster or agglomerate, leading to uneven distribution and creating weak spots within the composite. These clusters can act as stress concentrators, reducing the overall hardness. Increased reinforcement also introduces porosity or micro-voids in the composite, which can adversely affect the mechanical properties. There might be a saturation point beyond which adding more reinforcement does not lead to proportional increases in hardness due to the aforementioned factors.

3.3 Wear Rate Analysis

Table 5 illustrates wear rate data for two different samples (CS1 and CS3) at varying loads (10 N, 20 N, and 30 N). CS1 is the unreinforced Al-6063 sample, while CS3 is the Al-6063 composite reinforced with 4% Y_2O_3.

Table 5 Wear Rate of fabricated specimens.

Input	*Wear Rate (mm³/min)*		
Load (N)	10	20	30
CS1	6.27×10^{-4}	7.44×10^{-4}	9.16×10^{-4}
CS3	4.13×10^{-4}	4.87×10^{-4}	5.76×10^{-4}

The wear rate data presented in the table illustrates the impact of Y_2O_3 reinforcement on the wear resistance of Al-6063 composites under different load conditions. For the unreinforced Al-6063 sample (CS1), the wear rate increases steadily with the applied load, from 6.27×10^{-4} mm³/min at 10 N to 9.16×10^{-4}mm³/min at 30 N. This trend is typical for softer materials like aluminum alloys, where higher loads lead to greater material deformation and increased friction, resulting in a higher material removal rate. The wear mechanism in this case is likely a combination of abrasive and adhesive wear, with the material gradually being worn away by hard asperities or due to surface interactions that intensify with increasing load.

In contrast, the Al-6063 composite reinforced with 4% Y_2O_3 (CS3) exhibits significantly lower wear rates at all load levels, indicating enhanced wear resistance. At 10 N, the wear rate for CS3 is 4.13×10^{-4} mm³/min, which is considerably lower than that of CS1. This improvement continues under higher loads, with CS3 showing a wear rate of 4.87×10^{-4} mm³/min at 20 N and 5.76×10^{-4} mm³/min at 30 N. The presence of Y_2O_3 particles in the composite matrix plays a critical role in this enhanced wear resistance. These hard particles act as barriers to wear, reducing the extent of material removal by resisting abrasive forces and limiting adhesive wear interactions.

The reinforcement of Al-6063 with Y_2O_3 leads to a more uniform distribution of load across the composite material, reducing the localized stress and deformation of the softer aluminum matrix. This results in a surface that is better able to withstand higher loads without significant wear. The data suggests that the composite's wear resistance improves as a direct consequence of the Y_2O_3 reinforcement, which effectively mitigates the wear mechanisms typically observed in unreinforced aluminum alloys. Overall, the use of Y_2O_3 as a reinforcing agent significantly enhances the material's performance under varying load conditions, making the composite a more suitable candidate for applications requiring high wear resistance.

4. Conclusions

In this study, the conclusions were drawn based on the analysis of Al-6063 composites reinforced with Y_2O_3 particles, focusing on the impact of reinforcement on hardness and wear resistance. The findings highlight the optimal conditions for achieving improved mechanical properties and wear performance, aligning with the study's objectives as mentioned below:

1. Reinforcing Al-6063 with Y_2O_3 particles significantly increases the composite's hardness, with the highest hardness achieved at a 4% Y_2O_3 content.
2. The addition of Y_2O_3 enhances the wear resistance of Al-6063 composites, with the reinforced samples showing lower wear rates compared to the unreinforced alloy across all tested load conditions.
3. The composite with 4% Y_2O_3 exhibited the best wear resistance, indicating an optimal concentration for maximizing wear performance.
4. The improved wear resistance is due to the hard Y_2O_3 particles acting as barriers to wear mechanisms, particularly by resisting abrasive and adhesive forces.
5. Higher load conditions further demonstrate the effectiveness of Y_2O_3 reinforcement, with the composite maintaining superior wear resistance compared to the unreinforced alloy.
6. The study confirms that 4% Y_2O_3 is the optimal reinforcement content, providing the best balance between increased hardness and enhanced wear resistance in Al-6063 composites.

References

Kumar, D. (2023). Qualitative and quantitative interdependence of physical and mechanical properties of stir-casted hybrid aluminum composites. 51(6), 14–23. https://doi.org/10.18149/MPM.5162023.

Kumar, D., Singh, S. and Angra, S. (2024). Synergistic Effects of Graphene and Ceria Nanoparticulates on Microstructure and Mechanical Behavior of Stir-Cast Hybrid Aluminum Composite. *Transactions of the Indian Institute of Metals*, (0123456789). https://doi.org/10.1007/s12666-024-03368-y.

Kumar, D., Singh, S. and Angra S. (2023). Morphology and Corrosion Behavior of Stir-Cast Al6061-CeO 2 Nanocomposite Immersed in NaCl and H 2 So 4 Solutions. *EVERGREEN Joint Journal of Novel Carbon Resource Sciences & Green Asia Strategy*, 10(01), 94–104.

Kumar, D., Singh, S. and Angra, S. (2022). Effect of reinforcements on mechanical and tribological behavior of magnesium-based composites: A review. 50(3), 439–458. https://doi.org/10.18149/MPM.5032022.

Kumar, D., Angra, S. and Singh, S. (2023, May). Synthesis and characterization of DOE-based stir-cast hybrid aluminum composite reinforced with graphene nanoplatelets and cerium oxide. https://doi.org/10.1108/AEAT-04-2023-0104.

Kumar, D., Angra, S, and Singh, S. (2023). High-temperature dry sliding wear behavior of hybrid aluminum composite reinforced with ceria and graphene nanoparticles. Engineering Failure Analysis, 151(May), 107426. https://doi.org/10.1016/j.engfailanal.2023.107426.

Kumar, D., Angra, S. and Singh, S. (2023). Investigation of Corrosion Behavior of Stir-Cast Hybrid Aluminum Composite Reinforced with CeO_2 and GNPs Nanoparticles. 59(6), 1210–1218. https://doi.org/10.1134/S2070205123701186.

Kumar, D., Singh, S. and Angra S. (2023, Sept.). Synergistic corrosion protection of stir-cast hybrid aluminum composites reinforcing CeO_2 and GNPs nano-particulates. https://doi.org/10.1108/AEAT-03-2023-0067.

Kumar, D. and Kumar, P. (2024). Microstructural, microhardness, and electrical conductivity analysis of AD31T alloy processed by friction stir processing. https://doi.org/10.1108/MMMS-01-2024-0025.

Kumar, D. and Singh S. (2024, Jan.). Enhancing friction and wear performance in hybrid aluminum composites through grey relational analysis. *Research on Engineering Structures and Materials*.

Kumar, H., Kumar, V., Kumar, D. and Singh S. (2023). Wear Behavior of Friction Stir Processed Copper-Cerium Oxide Surface Composites. *EVERGREEN Joint Journal of Novel Carbon Resource Sciences & Green Asia Strategy*, 10(01), 78–84.

Kumar, D. and Singh S. (2024). Enhancing friction and wear performance in hybrid aluminum composites through grey relational analysis. *Research on Engineering Structures and Materials*, *x*(xxxx), 1–14. https://doi.org/10.17515/resm2024.05ma1012tn.

Kumar, S., Kumar, P., Tiwari, M., Panwar, V. and Kumar, D. (2024, Aug.). A comprehensive overview of advancements, applications, and impact of supercritical fluid natural circulation loops. *Annals of Nuclear Energy*, 211, 110971. https://doi.org/10.1016/j.anucene.2024.110971.

Kumar, D., Kumar, D., Gupta, A., Kumar, P. and Chauhan A. (2025). Enhancing Mechanical and Chemical Performance of Hybrid Polymer Composites : A Study on Oil Palm and Jute Fiber Reinforcement Enhancing Mechanical and Chemical Performance of Hybrid Polymer Composites : A Study on Oil Palm and Jute Fiber Reinforcement. 12(1), 453–459.

Kumar, D., Chalisgaonkar, R., Kumar, V. and Rai, S.K. (2024). A critical review of RAM methodology: Analysis and performance evaluation. *Reliability Theory and Applications*, 19(4), 83–89.

Kumar, P., Sharma, V.K. and Kumar D. (2024). Availability analysis for identification of critical factor of a thermal power plant. *Reliability: Theory and Applications*, 1(3), 717–724.

Kumar, D., Kumar, P. and Singh, S. (2025). Finite-element analysis of aluminum matrix composites reinforced with graphene nanoplatelets and cerium oxides for gear applications. In: *Proceedings of the Institution of Mechanical Engineers, Part E: Journal of Process Mechanical Engineering*, 09544089251347741.

Kumarm P., Sharma, V. & Dinesh Kumar, S.A. (2023). Morphology and Mechanical Behavior of Friction Stirred Aluminum Surface Composite Reinforced with Graphene. *EVERGREEN Joint Journal of Novel Carbon Resource Sciences and Green Asia Strategy*, 10(01), 105–110.

Patel, D.K. and Sharma, V.K. (2025). Enhancing hydrophobicity and anti-corrosion properties of Al-6061Aluminum hybrid composites through trace additions of rare earth oxides. https://doi.org/10.1177/09544089251317381.

Singh, S., Kumar, P. and Jain, S.K. (2013). An experimental and numerical investigation of mechanical properties of glass fiber reinforced epoxy composites. *Adv. Mat. Lett., 4.7* 567–572.

Singh, S., Surjit Angra, D.K. (2022). Optimization of Hardness and Tensile Strength of Stir Cast Hybrid Aluminum Metal Matrix Composite Using Grey Relational Analysis. In: *Proceedings of the ICCMSO Conference in Thailand*, pp. 132–136.

Shah, S., Kumar, P., Panda, S.K. and Kumar, S. (2019). Mixed-mode thermo elastic delamination fracture behavior of composite skin stiffener containing interface delamination. *Journal of Materials Research and Technology*, 8(6), 5941–5949.

Sharma, V.K., Kumar, P., Akhai, S., Kumar, V. and Joshi, R.S. (2024). Corrosion inhibition analysis on cerium induced hydrophobic surface of Al-6061/SiC/Al2O3 hybrid composites In: *Proceedings of the Institution of Mechanical Engineers, Part B: Journal of Engineering Manufacture*, 09544054231222048.

Sharma, V.K., Kumar, P., Akai, S., Kumar, V. and Joshi, R.S. (2023). Analyzing the tribological and mechanical performance of Al-6061 with rare earth oxides: An experimental analysis. In: *Proceedings of the Institution of Mechanical Engineers, Part E: Journal of Process Mechanical Engineering*, 09544089231160003.

Sharma, V.K., Kumar, V., Joshi, R.S., Jain, S. and Kumar, P. (2021). A Review of Recent Research on Rare Earth Particulate Composite Materials and Structures with Their Applications. *Transactions of the Indian Institute of Metals*, 1–13.

Index

About the Editors

Vipin Kumar Sharma has a Doctoral in Non-Traditional Machining Methods from Thapar Institute of Engineering and Technology, Patiala Punjab, India. He is currently working as a Professor and Head of the Department of Mechanical Engineering at Meerut Institute of Engineering and Technology, Meerut (UP), India. He has more than 15 years of teaching and research experience related to mechanical engineering. Dr. Sharma has published 20 research papers in reputed journals with an SCI indexing and 10 in major conferences. Dr. Sharma has received several academic distinctions, including the Best Paper Award (500 Euros) for his publication titled "A Review of Recent Research on Rare Earth Particulate Composite Materials and Structures with Their Applications," published in Transactions of the Indian Institute of Metals (Vol. 74, 2021)-recognized as one of the two best papers of the year. Dr. Sharma visited China for the presentation of his paper in the METADOR conference in the year 2019. His research interests include hybrid composites with Rare Earth Metals, Casting Methods, Optimization techniques and Design of experiments etc. Dr. Sharma is also a reviewer of many reputed journals including Journal of Materials Research Express and Advances in Materials Science. Currently, he is working on the fatigue analysis of hybrid composites reinforced with rare earth

Mohamed Arezki Mellal is a Professor at the Department of Mechanical Engineering, Faculty of Technology, M'hamed Bougara University, Boumerdès, Algeria. He was a Visiting Scholar at the Department of Mechanical Engineering, University of Maryland, USA and at some other universities. Prof. Mellal was in Standford University's Top 2% of Scientists Worldwide in 2022 and 1st by the 47th National University Commission for the Professorship in Mechanical Engineering in 2022. He was also involved in the international project "Algerian National Laboratory for Maintenance Education," in conjunction with the European Union (Erasmus+). He has supervised numerous dissertations, published in various SCIE journals, and conference proceedings. Prof. Mellal has edited 9 books and published 13 book chapters. He was an Invited Keynote Speaker at various conferences, and served as a Session Chair for the 2nd and 3rd International Conference on System Reliability and Safety of ICSRS 2017 and 2018, held in Milan (Italy) and Barcelona (Spain), respectively. He also chaired sessions at major conferences. Prof. Mellal was recognized as the Best Reviewer for the International Conference on Vision, Image and Signal Processing (ICVISP) in 2017 and 2018, held in Osaka (Japan) and Las Vegas (USA), respectively. He is an Associate Editor for the journal Frontiers in Mechanical Engineering (Topic: Mechatronics) – IF=2.3. He has served as a committee member for over 80 international conferences and acted as a Guest Editor for journal special sessions. He continues to serve as a regular reviewer for more than 30 SCIE-indexed journals. His research interests include the implementation of AI for solving engineering problems.

For Product Safety Concerns and Information please contact our EU representative GPSR@taylorandfrancis.com Taylor & Francis Verlag GmbH, Kaufingerstraße 24, 80331 München, Germany